GENETICS AND SOCIETY

GENETICS AND SOCIETY

Edited by
JACK B. BRESLER

ADDISON-WESLEY PUBLISHING COMPANY
Reading, Massachusetts · Menlo Park, California
London · Amsterdam · Don Mills, Ontario · Sydney

This book is in the

ADDISON–WESLEY SERIES IN THE LIFE SCIENCES

Consulting Editor

JOHNS W. HOPKINS III

 Printed in the United States of America. Published simultaneously in Canada. Library of Congress Catalog Card No. 72-2650.

ISBN 0-201-00600-6
BCDEFGHIJK-AL-7987654

This book is dedicated to the memory of
Isidore and Anna Bresler

Preface

How does the human gene pool affect human society and how does human society affect the gene pool? That is the subject of this book. The dominant themes throughout the volume are human behavior, human society, and human genetics.

Different kinds of papers are presented in this volume. Some are research papers and some are review papers on a specific topic. One is a review paper dealing with one man's work and thought. Another paper deals with a computer simulation. Still another deals with marriage laws of various states. A few papers provide concise data and are "conclusive" while others in the volume are frankly inconclusive. However, the manner and methodology in which hereditary problems are approached is important for their selection. Hence, there is a "mix" of different styles, different papers, different intents, different perspectives, and different conclusions. It is important for the newcomer to this field of study to note the varieties of format from which we obtain information on *Genetics and Society.*

The reader should know how the papers were selected. I have assumed that those reading this book have had some introduction to elementary genetics. This introduction could have come from a major section in a general biology, botany, or zoology course, could have come from taking an introductory course in genetics, or from extensive general reading of elementary genetics. I would recommend a very fine general introductory book dealing with human genetics (McKusick, Victor A., 1969. *Human Genetics,* Prentice Hall).

I have attempted to present many areas of human genetics and society in the limited amount of space available. Some fine papers in human genetics dealing with society were simply too long to include.

Throughout the volume, emphasis has been placed on population characteristics rather than single or individual characteristics. One problem intrudes here. Most reports on population genetics and demography contain some difficult passages of mathematical and statistical analyses. It has been my experience that many students do not have sufficient background in these forms of analyses. Hence, I had the dilemma of selecting papers in certain important areas which did not require knowledge of these more advanced mathematical and statistical requirements. Accordingly, the selection of papers has been modified to take this limitation into account. Obviously, I do encourage additional mathematical and statistical studies as proper preparation for further work in the field.

I also selected only those papers published in the most recent years. Information is accumulating rapidly in human genetics, demography, and behavior.

Thus, in selecting papers, I had to work under certain constraints: that the paper not be too long, that the paper not have too much advanced mathematics and statistics, that it deal with humans, and that it had been published in recent years. With these constraints as a sieve, certain deficiencies have arisen. For example, an appropriate introductory paper on assortative matings is lacking. However, this important concept is found in a number of papers throughout the book. The reader should receive a satisfactory review of assortative matings after reading those papers.

Each section will also have suggested Additional Readings with a few comments on the recommended papers. Also at the end of each section of the book there are Questions about themes of that particular section. More frequently the questions will relate to broad issues. The "answers" to the questions will be found in many sources of which only one will be this book.

Inevitably the question of commenting on the so-called Jensen report (1969) in the Harvard Educational Review and the recent Herrnstein Atlantic magazine article (1971) will arise. These reports were published during the formative stages of this volume. It was my original intention to prepare students and others to understand the issues of genetics and society. After reading the papers by Jensen and Herrnstein, I have become all the more convinced that my original intent to further the understanding of genetics and society was correct. Therefore, I have tried to include many papers on genetic patterns, human development, environmental influences, and intelligence which represent preparation in understanding the import of the issues raised by Jensen and Herrnstein. It is one thing to use polemics and rhetoric to support or condemn these papers, it is another to understand the genetic and environmental methodologies so necessary for an honest critique. See also a commendable analysis of these papers by Scarr-Salapetek (1971) in *Science.*

Section A has a single paper on the contributions of H. J. Muller. Sections B, C, and D form an aggregate of papers concerned with human development and interactions between heredity and environment. Sections E and F deal with the characteristics of ethnic groups and what happens with matings between ethnic and racial stock. Section G contains papers on the human restructuring of his own species.

The commentary on each paper is minimal, by intent. The selection process of the papers has been rigorous. The reader with some background in genetics, a medical dictionary, and possibly an introductory statistics text readily available should have few problems.

References Cited

Jensen, Arthur R., 1969, Environment, heredity, and intelligence. *Harvard Educational Review,* Cambridge, Mass., IV, 248 pages.

Herrnstein, Richard, 1971, I. Q. *Atlantic,* September, **228**:44-64.

Scarr, Sandra Salapatek, 1971, Unknowns in the I. Q. equation. *Science,* **174**:1223-1228.

Newton, Massachusetts J. B. B.
November 1972

Contents

A **The Contributions of H. J. Muller** **1**

Introduction*

1 Science and Society in the Eugenic Thought of H. J. Muller . . 3

Garland E. Allen

B **Basic Genetic Patterns** **25**

Introduction

2 Kinship of Smenkhkare and Tutankhamen Demonstrated Serologically 28

R. G. Harrison, R. C. Connolly, and A. Abdalla

3 Reactions to Coffee and Alcohol in Monozygotic Twins 30

K. Abe

4 The Incidence of Cuna Moon-Child Albinos 37

Clyde Keeler

5 PTC Tasting in Two Social Groups 48

Christy G. Turner II

C **Human Chromosomes and Antisocial Behavior** **51**

Introduction

6 Incidence of Gross Chromosomal Errors Among Tall, Criminal American Males 54

Mary A. Telfer, David Baker, Gerald R. Clark, and Claude E. Richardson

*All introductions by J. B. Bresler

7 Human Chromosome Abnormalities as Related to Physical and Mental Dysfunction 57
John H. Heller

D Genetics and Early Human Development **83**

Introduction

8 Delayed Radiation Effects in Atomic-bomb Survivors 86
Robert W. Miller

9 An Investigation of the Difference in Measured Intelligence Between Twins and Single Births 102
R. G. Record, Thomas McKeown, and J. H. Edwards

10 Genetics and Sociology: A Reconsideration 117
Bruce K. Eckland

E Genetics and Ethnic Characteristics **159**

Introduction

11 Gene-frequencies in Jews 162
Chaim Sheba

12 Welshness and Fertility 166
David J. B. Ashley

13 Behavioural Differences between Chinese-American and European-American Newborns 174
D. G. Freedman and Nina Chinn Freedman

F Origins of Human Matings: Genetic and Social Issues **179**

Introduction

14 The Founder Effect and Deleterious Genes 182
Frank B. Livingstone

15 Some Genetic Aspects of Plantation Slavery 191
Nathaniel Weyl

16 Outcrossings in Caucasians and Fetal Loss 200
Jack B. Bresler

G Medicine and Law Consider Human Genetics **215**

Introduction

17 Genetic Counseling and the Physician 217

Henry T. Lynch, Gabriel M. Mulcahy, and Anne J. Krush

18 Genetics and Laws Prohibiting Marriage in the United States 228

Michael G. Farrow and Richard C. Juberg

H The Consequences of Social Selection and Genetic Engineering **241**

Introduction

19 Some Possible Genetic Implications of Carthaginian Child Sacrifice 244

Nathaniel Weyl

20 Freeze Preservation of Human Sperm 254

S. J. Behrman and D. R. Ackerman

21 Should Man Control His Genetic Future? 268

Donald Huisingh

The Final Question 281

A.
The Contributions of H. J. Muller

The eugenic contributions of H. J. Muller are significant and permeate all discussions which relate genetics to society. Many references to him will be made in papers found in this book. It is therefore most appropriate for us to understand what this man has written and placed before us The influences which have acted on Muller as well as the chronological progression of his thinking on science and culture are the subjects in Allen's review.

Although Muller wrote much about eugenics, the eugenics movement was in existence before his birth and is now about 100 years old. Shortly after the eugenics movement began, two complementary approaches diverged. The first eugenic approach was the positive encouragement of those with favorable genetic constitutions to reproduce. The second approach was the discouragement, sometimes by law, of those people with genetic constitutions believed to be undesirable. At times the second approach has been used in horrendous proportions—as, for example, the Nazi notion of inferiority and subsequent extermination.

H. J. Muller received the Nobel prize for his research involving radiation effects on hereditary materials. Radiation probably represents one of the earliest man-made attempts to use an environmental factor to change a gene pool constitution of an organism. Later in this book, a paper (No. 8), "Delayed radiation effects in atomic bomb survivors," shows how radiation from an atomic bomb has affected a human population sample. Hence from Muller's early years to the present, a span of approximately 50 years, there has been an uninterrupted line of research dealing with radiation and its effects on gene pools.

Today, man-made radiation from all sources possibly represents the greatest environmental factor introducing deleterious genes into the human population. The immense contribution by Muller on the devastating effects of radiation appears to have little effect on human thought and action.

Additional Readings

Winkelstein, Warren Jr., and David Sackett, 1968. Eugenics and genetic equilibrium. *Clinical Pediatrics,* 7:2-3. A short paper relating eugenics to new community health services.

Hirsch, Jerry, 1970. Behavior—genetic analysis and its biosocial consequences. *Seminars in Psychiatry,* **2**:89-105. An excellent overview of behavioral genetics with short statements on the alleged genetic inferiorities of Jews and blacks.

Questions

What is Social Darwinism?

Which ethnic groups have at one time or another been judged genetically inferior? Superior? For what reasons?

1. Science and Society in the Eugenic Thought of H. J. Muller

GARLAND E. ALLEN

INTRODUCTION

My task at this symposium is to discuss the impact of biology on culture over time. I have taken this to mean the analysis, through historical methods, of the way biological ideas have influenced man's view of himself and his social organization. In general, this paper will deal with the development of Mendelian genetics in the 20th century. To prune the subject down to a more manageable size, I propose to focus on the growth of scientific theories of eugenics during this time, particularly in the work of Hermann Joseph Muller (1890–1967).

There are many ways in which the rapid growth of Mendelian theory during the first two decades of the 20th century has influenced our culture. Deducing patterns of heredity has made possible human genetic prognostication, and thus a more thorough understanding of the nature and treatment of hereditary disease. Extremely rapid progress in the science of plant and animal breeding (for example, in developing the modern-day strains of wheat) could not have occurred without an understanding of the laws of heredity. In the use of genetic information to improve the quality of the human hereditary stock, however, modern genetics has reached its potentially most powerful impact on modern culture.

For the historian, Muller's investigations in genetics and in the problems of planned human breeding present an ample opportunity to study the interaction between scientific and social views. In Muller's mind, the two were inextricably connected. His ideas on the importance of directing human evolution motivated his thorough and far-reaching work in genetics. Conversely, his findings about gene action and mutation profoundly influenced his views as to how, and under what conditions, society could modify its breeding habits. While studying in detail the interaction of such ideas in the work of a single individual, the historian can hope to draw some tentative conclusions which have wider applicability. In this case, Muller's work may be used to illustrate: (1) the general interaction of social and scientific ideas which form a part

Reprinted by permission from *BioScience,* 1970, Volume 20, 346-352.

of the cultural *milieu* of any epoch; and (2) how, in this one case, biological ideas have influenced and to a large extent determined a man's ideas of social and ethical reform.

In a paper published in *Science,* T. M. Sonneborn (1968) discussed at some length Muller's ideas on improving the human genetic stock. The present paper is an attempt both to expand upon and give a different slant to Professor Sonneborn's discussion. First, I would like to relate more explicitly Muller's genetic work to his views of human evolution and the structure of human society. While Professor Sonneborn treated these relationships to some extent, they were not the major focus of his paper. Second, I would like to set Muller's work in the broad context of the development of eugenic ideas in the later 19th and early 20th centuries.[1]

THE EARLY HISTORY OF EUGENICS IN AMERICA

Growing directly out of evolutionary ideas first proposed by Darwin in *The Origin of Species,* the eugenics movement was founded in England in the 1870's and 1880's by Francis Galton (1822–1911).[2] Galton, Darwin's cousin, was intrigued by the conception of human evolution hinted at in *The Origin of Species.* He spent much of his life collecting data about human variation and studying what he thought were hereditary changes in human population. The term "eugenics" itself was first coined by Galton in a book of 1883 titled, *Inquiries into Human Faculty.* It was, as he wrote: "The study of the agencies under social control that may improve or impair the racial qualities of future generations either physically or mentally" (Galton, n.d.). No stronghold of ideas about human equality, Galton showed—through many of his studies—that superior hereditary lines tend to be diluted by crossing with inferior lines, and that human offspring tends to be more like the average between their parental strains than like either parent. Coupled with his knowledge that inferior genetic types (judged mainly by intelligence) tended to reproduce more rapidly than superior ones, Galton's genetic ideas indicated that unless strong measures were taken, the human evolutionary course could only be downhill. To educate the public about this problem as well as to investigate means by which it could be prevented, Galton spent the latter years of his life in the founding and organization of the National Eugenics Laboratory in England.

While Galton's influence in his native country and on the continent was reasonably strong, the eugenics movement found its most fertile ground for development in America (Hofstadter, 1955). Haller has divided the eugenics movement, both in this country and abroad, into three periods:

> *1870–1905:* This period saw the development of strong hereditarian attitudes among those concerned with human social problems. Poverty, feeble-mindedness, and insanity were all judged to be ultimately the result of hereditary defect.
>
> *1905–1930:* During this period reinforcement grew for the idea that all human weaknesses were the result of poor heredity. At the same time, especially in America, immigration was seen as detrimental to the American hereditary stock. Particularly, the large influx of peoples from southern and eastern Europe led to the promulgation of severe immigration restrictions.
>
> *1930–to present:* During this period the eugenic movement lost some of its momentum and gradually went downhill. New studies in areas such as psychiatry, heredity, anthropology, and mental testing undercut the scientific basis on which the earlier movement had rested. All of these studies combined to suggest that human heredity was much more complex than previously imagined and that environment had a much more important role in determining an individual's makeup than had been imagined previously. At the same time, an example of the extremes to which ideas of selective breeding and control of human evolution could be perverted was visible in Hitler's programs to insure Teutonic mastery of Europe (Haller, 1963).

In America, the eugenics movement took firm root by the turn of the century, making its presence felt in a variety of ways. For example, The American Breeders Association, founded in 1901 as a forum of interchange between professional breeders and theoretical geneticists, had established a eugenics subsection as early as 1903. And further, by 1910, a group of eugenists, with the financial assistance of Mrs. E. H. Harriman, founded the Eugenics Record Office at Cold Spring Harbor, New York. This organization became a laboratory for collecting statistical information on human heredity and variation, while also serving as a soap box from which colorful propaganda could be circulated (Hofstadter, 1955). On a more practical level, the State of Indiana in 1907 adopted a sterilization law for the feeble-minded and for inveterate criminals, and by 1915 twelve other states had followed suit.[3] It is clear that by 1920 a high degree of interest in eugenics as a means of solving many of the ills of society had entered into the worlds of medicine (Barker, 1910), academics, and social legislation (Hofstadter, 1955).

Interest in eugenics in the late 19th and early 20th centuries seems to have been promoted by two general factors: (1) changing social conditions; and (2) the impact of two important biological discoveries—Weismann's Germ Plasm Theory, and Mendel's laws of heredity. Consider first social conditions.

Rapid urbanization in the post-Civil War period had created in America's major cities vast slums which housed a large section of the diseased and demented—the derelicts of society. Such slums provided a vivid indication to the average man of the depths to which the human condition could sink. Many eugenicists argued that one way to prevent this problem was to prevent the future slumdwellers (thought of as synonymous with the genetically inferior) from being born. At the same time, and perhaps stimulated by recognition of these deplorable conditions, there was a growing interest in philanthropy: the increase of gifts to charities, hospitals, and the development of schools of social work. Equally as important, however, to the changing social scene in the United States was the influx of large immigrant groups from Europe. With their different languages and customs, the immigrants were naturally slow and halting in their discourse with native Americans. Assuming, as Hofstadter puts it, that a glib command of English was a natural criterion of intellectual capacity, many Americans argued that these immigrants were of inferior mentality, and thus were lowering the standard native intelligence (Hofstadter, 1955). Interest was thus stimulated in eugenical arguments for restricting immigration and for controlling the breeding of certain immigrant groups thought to be biologically inferior.

Of the two biological ideas which, during this period, gave impetus to the hereditarian approach in sociology, the theory of the continuity of the germ plasm, proposed by August Weismann in 1885,[4] was the most immediately influential. Weismann's work emphasized the genetic continuity which exists between parent and offspring by pointing out that, from the first cell-division onward in a developing embryo, the cells which later form germinal tissue are morphologically distinct from the cells which form the rest of the body. Changes in body cells during an animal's life could not be transmitted to the germ cells and thus not to the offspring. Weismann used this conception to argue against the then popular theory of the inheritance of acquired characteristics. The general effects of Weismann's idea, both inside and outside the biological community, was to emphasize the role of heredity and to de-emphasize that of environment in determining the characteristics of an organism.

This idea was strengthened in the decade after 1900 by the rediscovery of Mendel's important work of 1865. Mendel's "factors" not only proved to be real entities (as genes located on chromosomes) whose patterns of inheritance could be followed in a predictable way, but they were also discrete hereditary units which were not modifiable by other genes or agents from the external environment. Together, the theories of Mendel and Weismann provided scientific sanction for the hereditarian approach to social problems.

The eugenics movement gained a foothold in America at a time when the doctrines of social Darwinism still held strong sway. As an extension to human society of the theory of natural selection, social Darwinism in its purest form maintained that those individuals who were economically, socially, or politically successful were the biologically more fit, while those who were unsuccessful in these endeavors were the less fit. The rich occupied privileged positions, not because of their own greed or avarice but because it was inevitable that, as the most favorably endowed organisms, they must come out on top. In accumulating great wealth, or in exploiting human beings, it was argued that men were only obeying a natural, cosmic law. The working classes were where they were because they could not compete successfully with more fit individuals, and thus occupied their natural position in society.

In America, particularly through the works of Herbert Spencer (1820–1903), social Darwinism glorified the individual and emphasized the function of competition in the advancement of society. Somewhat later, this idea was extended in a slightly different form as the imperialist doctrine of Manifest Destiny on the one hand and the ethnic doctrine of Anglo-Saxon superiority on the other (Hofstadter, 1955). American society saw in natural selection its own image. In politics and business especially, competition and fitness became the dominant ethic. It was assumed that the unfit would be selected against, and that society as a whole could only advance to higher levels of adaptation and success—physically and morally—by the interplay of these natural forces.

The political mood of social Darwinism was thus strongly conservative. Like the proponents of laissez-faire economics, social Darwinists maintained that any tampering with universal laws was "unnatural" and doomed to ultimate failure. Consequently, reforms aimed at counteracting the selection process in human society were strongly opposed by both business and political leaders. Federal or state legislation on such matters as unemployment, collective bargaining, wage and hour laws, or economic regulation met constant opposition by those who were, as might be expected, the individuals who were in a position to declare themselves "most fit" for guiding society.

There were, of course, dissenters, and the social Darwinist movement was never a single, monolithic idea. Its ultimate decline as a viable philosophy came both through social and intellectual channels. The deplorable exploitation of masses of human beings which was blatantly justified as "natural" came to be intolerable to a large segment of the population. At the same time, critics pointed out the circular logic of the social Darwinism philosophy: It was claimed that the "fit" were the ones who survived, yet "fitness" was always defined as being able to survive. Thus, by the turn of the century, social Darwinism was no longer as openly avowed a doctrine as it had been in the 1870's or 1880's. Yet, the conservative turn of mind which social Darwinism attracted was still a dominant force and, in many cases, was attracted to the newer ideas of eugenics.

Like the older social Darwinism, eugenics emphasized heredity over environment in determining human character. Unconsciously, eugenicists, like the social Darwinists, identified the "fit" in a human population with the upper classes and "unfit" with the lower. It was argued that the poor are held down by their biological deficiencies rather than by environmental conditions. For example, in 1911, no less eminent a person than David Starr Jordan, probably the foremost ichthyologist in America and then president of Stanford University, stated that poverty, dirt, and crime could be ascribed to poor human heredity, adding that, "it is not the strength of the strong but the weakness of the weak which engenders exploitation and tyranny" (Jordan, 1911). Of course not all social theory at the turn of the century was hereditarian in attitude. Professional sociologists stressed strongly the role of the environment and tended to accept eugenic ideas with some hesitation. Nevertheless, where eugenic ideas found great favor, it was on many of the same arguments which had given social Darwinism its great popularity in the 1870's and 1880's—the view that the structure of society was determined by hereditary rather than environmental factors.

As might be expected, eugenicists were, for the most part, opposed to socialism and its attempts to work social wonders by changed environmental conditions. Like the social Darwinists, eugenicists tended to take a laissez-faire attitude toward social change. They opposed minimum wage laws and the formation of labor unions on grounds that both would encourage the propagation of the more unfit members of society. However, the eugenicists differed from the social Darwinists in one important respect. Eugenics sought a social end through individual cooperation within society. Social Darwinism, on the other hand, emphasized competition and lack of cooperation between individuals. Thus, by comparison, the eugenics movement had a more progressive flavor than the ultrareactionary views which characterized most of the social Darwinist movement. Eugenicists saw that social change must involve concerted effort on the part of the members of society, and this was generally more than most social Darwinists would admit.

The eugenics movement reached its peak of general interest by about 1915, with the publication of a number of textbooks and semipopular accounts. The works of Major Robert Darwin, Paul Popenoe, and Roswell H. Johnson, among others, all brought the eugenics platform before the lay reader. At the same time, C. B. Davenport, the scientific leader of American eugenics, was publishing volumes of data —measurements and statistics of all sorts—from his laboratory at Cold Spring Harbor. Davenport (1919) measured every conceivable aspect of living, and even historical figures, in an attempt to derive general laws about the nature of human hereditary deficiencies. Although Davenport's scientific work ultimately led to far less than he had hoped, it stimulated considerable interest in the study of human heredity and evolution.

MULLER'S EARLY INTEREST IN EUGENICS

Born in New York City in 1890, H. J. Muller graduated from Columbia College in 1910, the same year that T. H. Morgan discovered the first mutant in the fruit fly *Drosophila.* Immediately after graduation, Muller went to work with Morgan's group and participated in the important investigations which led to the establishment of the chromosome theory of heredity (Morgan et al., 1915). In 1916, Muller received his Ph.D. from Columbia for a study of the interrelations of linked genes in *Drosophila.*

Muller's strong interest in science stemmed from his childhood and, according to his testimony, specifically from his father (Carlson, 1967). An interest in biology in general, and in the problem of heredity in particular, came in 1908 when Muller, then a sophomore, read R. H. Lock's *Recent Progress in the Study of Heredity, Variation and Evolution* in preparation for a course in cytology by E. B. Wilson.[5] Lock's book was important because it related Mendelian genetics to evolution. Lock showed that (1) Mendel's work provided the one method of understanding heredity in any systematic way; (2) genes could be thought of as physical parts of chromosomes (including a suggestion of exchange between linearly arranged genes); and (3) genes could mutate and that such mutations could serve as the primary events of evolution.[6] In addition, Lock's book may have had a further influence which Muller did not specifically record. Chapter 10 of this book deals with problems of eugenics and the control of human evolution. As Morris points out, the ideas which Lock proposed in this chapter represent the main line of eugenic ideas which Muller was to propound over the next 30 years.[7]

How specific the influence of Lock's eugenic discussion was on Muller's thinking is difficult to establish. However, Muller does indicate in his autobiographical sketch that interest in controlling human evolution was the initial stimulus which directed his attention to genetics: "The original source of my interest in genetics had been my long-harbored idea of the control of the evolution of man by man himself. I had intentionally, however, devoted most of my efforts to the investigation of the general genetic basis, being convinced that this would provide a surer foundation and backing for a later attack on more specifically human problems. Only so could the necessary knowledge as well as the authority be obtained" (Sonneborn, 1968).

Whichever may, in reality, have come first, it is apparent that Muller's interest in eugenics arose simultaneously with his interest in the new science of Mendelian genetics. A strong desire for improving human heredity directed Muller's attention to the study of genetics and evolution. At the same time, an understanding of how hereditary characters are passed on and how new variations arose required knowledge that could be obtained only from rigorous scientific investigation. The constant interplay between these two concerns formed the basis of Muller's professional life.

Muller's college and graduate school years coincided with the peak of activity in the American eugenics movement. Whether the abundance of eugenic literature which appeared around the turn of the century kindled or fanned the flames of Muller's interest, it is difficult to say. It would be surprising, however, if the widespread interest in eugenics did not act, if only catalytically, in determining his profound concern with human evolution.

The earliest formulation of Muller's eugenic ideas was given in 1910 at a meeting of the Peithologian Society, a student club at Columbia College. In this paper, titled "Revelations of Biology and their Significance," Muller raised the problem of the accumulation of genetic defects in the human population and how biology could prevent this. He suggested first that the way to eliminate the unfit is simply to keep them from being born, by various sterilization procedures.[8] But sterilization is only a negative process. It must be accompanied by positive steps, by which Muller meant specifically artificial insemination.[9] Any successful eugenic program required man to begin thinking *now* about the future. To Muller, humanitarianism in its broadest aspect included a deep concern for the future as well as the present condition of mankind. The urgency with which, in this early paper, Muller viewed the problem of man's action in behalf of his own hereditary future remained part of his approach to eugenic problems throughout his life.

Genetic improvement of the human race implied certain criteria by which the fit and the unfit could be judged. Muller himself in 1910 was not certain exactly how such judgments would be made to insure optimal evolution. He did suggest, however, that by the interaction of science, psychology, and sociology proper criteria could be developed: "Science, in the form especially of psychology and sociology, will discover what qualities are desirable for the most efficient cooperation and for the best enjoyment of life; and science, in the form especially of physiology and genetics . . . will discover what the elementary bases of these qualities are and how to procure them for man."[10]

To Muller, as early as 1910, planned genetic control could only be beneficial to mankind if it were carried out on a large scale and in a rational way. He saw eugenic control as one of the most important ways man could use his new technology and skills to transcend human weakness and misery. As he said, "Mankind has nothing real to lose but only to gain, by a process of evolution. . . . Only tradition is opposed to the plan, and our own stupidity and defective social nature . . . these prevalent notions will go as others have gone; but upward change will remain. With knowledge of the laws of nature comes power to manipulate them, and knowledge of life thus means perfection of man."[11] The youthful optimism and idealism with which the 20-year-old Muller spoke to his fellow students in the Peithologian Society never left him. The purpose of his life's work in science was to realize the ultimate aim of the perfection of man.

Although imbued with the eugenic spirit early in his life, Muller was prescient enough to realize that no answers could be given without the background of firm and rigorous scientific information. He thus began his studies on gene variability and mutation shortly after receiving his Ph.D. in 1916. The subsequent genetic investigations in which he engaged were devoted essentially to answering the questions of how genes are inherited, how they influence each other, and how mutations arise.

MULLER'S RADIATION STUDIES

Radiation as a cause of mutation has been studied to some extent in 1910 and 1911, both by T. H. Morgan and Jacques Loeb,[12] though quite independently. Muller may have picked up the idea here of using radiation to try and increase the number of mutations in living organisms, or it may have come slightly later in his joint work with his close friend Edgar Altenberg on the induction of mutations. At any rate, in 1916 Muller began working with X-radiation (which is easier to quantitate than radiation from radium samples). His purpose was to determine whether radiation caused mutation or not, and if so, what relation existed between artificially and naturally induced mutations. Mutation, as the term is used in Muller's work, refers to the same kinds of small but definite, heritable variations implied in our present-day use of the term. It is thus distinct from the large-scale "mutations" discussed by the Dutch plant physiologist, Hugo deVries, in the first decade of the 20th century (where a single mutation was supposed to produce a wholly new species).

Muller's radiation work can be conveniently divided into three periods: 1916–1921, 1921–1927, and 1927 through the 1950's. During the first period, Muller was unable to show conclusively that mutations were actually caused by high-energy radiation. But he did come to grips with two important and related problems. One was the understanding that rate of mutation rather than kind of mutation was the important problem for understanding evolution (Muller and Altenberg, 1919). The other was to calculate the natural mutation rate in *Drosophila.* Statistical evidence showed that approximately one out of 53 flies had a mutation on the X chromosome. Knowing that the X's form approximately one-fourth of the entire chromosome mass, and assuming that mutation rate for other chromosomes is about the same as for the X, Muller calculated that one out of every 13 flies has a mutation somewhere in its genome. In less than 100 fruit fly generations (about 4 years), every X chromosome in a *Drosophila* population would be expected to have a mutation. That this has not happened in nature was a testimony to the effectiveness of natural selection in weeding out deleterious mutations. The only mutations which showed up, then, were those which had a positive, or at least a not-too-negative, survival value. On this basis, Muller and Altenberg showed that all of the X chromosomes in a female would contain several lethal factors by the time she would be ready to reproduce and consequently none of her sons would be viable (Muller and Altenberg, 1919).[13]

Although there was no evidence at the time that the human mutation rate was necessarily the same as that for *Drosophila,* Muller was undoubtedly impressed by the enormous possibility for accumulation of defective genes in a population. As he pointed out in the case of *Drosophila,* were the effects of the natural mutation rate not greatly diminished by natural selection, the accumulation of lethal factors would inevitably bring about the elimination of a population after only a few generations.

The second stage of Muller's mutation studies (1921–27) brought forth his most brilliant and far-reaching results. During this time, he developed methods for accurately and quantitatively determining the effects of radiation on mutation by devising a stock of fruit flies (the so-called CIB stock), with three known markers on the X chromosome.[14] The importance of this method was that it allowed Muller to immediately identify, by phenotypic observation of the offspring of any cross, the number of mutations which had been induced by exposure to X-radiation. The uniqueness and convenience of this method made it possible to study large numbers of flies and to determine quantitatively the actual effect of radiation on mutation rate. From the CIB studies, Muller drew three important conclusions. First, he saw that artificially induced mutations were qualitatively of the same sort as natural mutations; second, it was apparent that most mutations were lethal; and third, breeding results showed that the transmission of nonlethal, induced mutations followed a strictly Mendelian pattern (Muller, 1927). It was for this work that Muller received the Cleveland Research Prize from the American Association for the Advancement of Science in 1927, and the Nobel Prize in Physiology and Medicine in 1946.

The third stage of Muller's mutation studies involved building upon the conclusions reached with the X-radiation of the second period. In 1927, for example, Muller had suggested that the relationship between mutation rate and dosage of X-radiation was proportional to the square root of the radiation absorbed. Further studies showed that dosage had a linear effect on mutation rate. They also showed that mutations were cumulative in the germ plasm of any genetic line. Determining the exact relationship between dosage and mutation rate was important not only because it helped to clarify the nature of mutation but also because it led to Muller's concern for human evolution. To Muller, anything which increased the mutation rate in man was pushing the species dangerously close to the brink of possible extinction through the accumulation of unwanted genes. Thus, it was extremely important to know what environmental factors (principally, amounts of high energy radiation to which humans were exposed) induced mutation and at what rate. Muller's eugenic concerns had led him to rigorous laboratory analysis of gene mutation. Conversely, his findings in the laboratory now led him back to an even greater concern for the genetic well-being of man.

MULLER'S REVISED EUGENIC THINKING

What Muller had asked for in 1910—sounder biological knowledge of hereditary and mutation—to some extent had been provided by 1930. In relating mutation rate to evolution, Muller could now argue from a factual point of view. Eugenics was, after all, merely the controlled evolution by man of his own genotype. In the early 1930's, Muller emphasized several points about mutation and evolution: (1) Man and all animals are subject to natural radiation even if not to artificial. Thus, mutations are constantly arising in any genetic stock (Muller, 1962). (2) Although most mutations are detrimental, some very few are advantageous to the organism. Evolution through natural selection exploits these few advantageous mutations (Muller, 1929; Muller and Patterson, 1930). (3) Muller could now apply these ideas to increasing rates of evolution in experimental animals, especially *Drosophila.* By selecting the few favorable variations, change in the phenotype of the population (and consequently its genotype) could be effected in almost any direction.

Muller was quick to point out, in 1934, the practical value of this point in the selective improvement of agricultural stock (Muller, 1962).

It was only a short step for Muller to go from the facts of mutation and selection in animals and plants to the problem of controlling human evolution. In human populations as in populations of other species, variations arise, are acted upon by one selective force or another, and are passed on to the next generation in greater or lesser proportions than other genes. The important difference between man and another species is that the former has become aware of the nature of genetic variations and can act to control his genetic destiny. As Muller wrote in 1935: "There is no permanent *status quo* in nature; all is the processes of adjustment and readjustment, or else eventual failure. But man is the first being yet evolved on earth which has the power to note this changefulness, and, if he will, to turn it to his own advantage, to work out genetic methods, eugenic ideas, yes, to invent new characteristics, organs, and biological systems that will work out to further the interests, the happiness, the glory of the God-like being whose meager foreshadowings we the present ailing creatures are" (Sonneborn, 1968).

To Muller, the processes which favored human genetic engineering were at our disposal. Man had only to be rational and far-sighted enough to put them into effect.

Muller's views were idealistic and, indeed, Utopian. As he foresaw, they did not have a chance of gaining acceptance in a capitalistic society, especially one which was, in the early 1930's, experiencing one of the worst financial depressions in its history. In general, the American social system seemed to Muller remarkably insensitive to human suffering and to the ideal of human betterment. It was not surprising that at this time he moved to the political left: toward socialism and eventually Marxism. Muller was convinced that his eugenic ideas could only flourish in a socialistic state. Under socialism, the individual ego would become less important than under capitalism, where individuality and competition are the basis of social and economic process. Under socialism, too, all members of society would work together for the common good, for concerted social action, as opposed to capitalism where each individual works primarily for himself. To Muller, capitalism was too predatory a social system—it thrived, in his terms, on exploitation of the inferior genotypes by the superior.

It was thus with enthusiasm that Muller, in 1933, while on a Guggenheim Fellowship at the Kaiser Wilhelm Institute in Berlin, accepted an invitation from N. I. Vavilov to become a senior member of the Institute for Genetics in Moscow. Muller was to establish his own laboratory in Leningrad and receive financial support from the Institute, as well as membership in the USSR Academy of Sciences. In Russia, he could not only continue his genetic research with radiation but also find a cultural and social *milieu* more congenial to his eugenic ideas than the United States. However, Muller was ultimately destined to find that Marxist socialism was, in fact, less congenial to both his scientific and social views than he had originally hoped. The rise to prominence of T. D. Lysenko and the subsequent domination of biological ideas by doctrinaire political theory brought forth Muller's full opposition. But Lysenko had Stalin's support, and on the advice of Vavilov, Muller left Russia by volunteering to serve in the Canadian Blood Unit during the Spanish Civil War (Carlson, 1968).

While in Russia, Muller had organized his growing eugenic ideas into a book titled, *Out of the Night: A Biologist's View of the Future* (1935). This book begins with the key idea that the purpose of life is to insure equal opportunity to all for material and social happiness. This means that both human environment and human genotypes must be developed in as optimal a direction as possible. The best genotypes cannot function well in an environment which restricts their freedom and potential of expression. At the same time, the most favorable environment cannot evoke optimal responses from a genetically inferior or incapable individual.

Muller was able to see in this relationship, however, the necessity of providing the right intellectual and social environment for the concept of eugenics itself to take hold. Thus, the first step in his eugenic program was to eliminate the present environmental condition of economic exploitation—which meant eliminate laissez-faire capitalism. To Muller, capitalism prevented cooperation among individuals for the good of the society as a whole. It is not surprising, then, that the ideal characteristics for which Muller would select were comradeliness and intelligence. By selecting for such traits (which Muller assumed without much evidence were inherited), it would ultimately be possible to eliminate much human suffering and such active horror as war.

Despite his considerable scientific achievements up to this time, Muller did not bring much technical detail to bear on his eugenics ideas in *Out of the Night.* Rather, this book represents the voice primarily of Muller the social theorist. It shows clearly the important relationship which he saw between the structure of society and the use by that society of available resources for its own improvement. Even more important, *Out of the Night* occupies a significant place in the history of eugenic writing. For the first time in the history of social thought Muller combined strong eugenic ideas with liberal (even radical) political and social philosophy. One of Muller's lasting contributions to the concept of eugenics was to partially liberate it from the conservative and reactionary philosophies of the late 19th and early 20th centuries which had used scientific knowledge to justify privilege and brutality.

THE NOBEL PRIZE AND MULLER'S LATER EUGENIC THOUGHT

In 1947 Muller was awarded the Nobel prize in medicine in recognition of his work on the artificial production of mutation by X-radiation. The awarding of the Nobel Prize not only increased Muller's stature among his colleagues and the public at large, it brought into sharp focus the timeliness of his scientific and sociological views. The honors and responsibilities which the award conferred, along with mounting public concern over the effect of radiation from atomic testing, gave Muller the public forum and the international prestige necessary to publicize forcibly his views about genetic defects and human evolution. In particular, Muller voiced considerable concern about the deleterious effects of radiation on the human gene pool, and the necessity for full measures of protection against radiation damage (Carlson, 1967). To this end, Muller served on a number of national and international commissions for the control and regulation of atomic energy.

However, radiation protection was, to Muller, only one part of the general eugenics program which he now aggressively promoted. As we have seen, he nurtured his eugenic ideas from his student days onward. By the 1950's he was able to put these ideas forth in a systematic and forceful way, using the full power of his technical knowledge and personal prestige. The period after World War II represents the final stage in the development of Muller's eugenic ideas.

Muller began his systematic thinking with the premise that modern medicine and society encourage unequal rates of survival in favor of those persons laden with mutated genes. By this he meant simply that medicine, by finding ways to keep individuals who carry deleterious genes alive, encourages the survival of those who are a burden to society and who more than likely will pass on their defects to future generations. To counteract this, Muller felt now, as before, that two complementary solutions existed: elimination of the unfavorable genes while at the same time increasing the number of favorable genes in the population.

Elimination of unfavorable genes could be brought about by preventing children from surviving who had obvious and observable genetic defects, and by control of reproduction in the genetically unfit through sterilization or contraception. Both of these methods raised some serious problems which Muller noted. Infanticide, even by knowledgeable medical personnel, is difficult for most members of society—any society—to accept morally. Sterilization is equally as difficult to accept, while contraception works only if all or most of the population practices it with regularity. Muller went on to point out that all of these measures were primarily negative: they worked at eliminating unfavorable genes, but did not provide means of significantly increasing the number of favorable genes.

Far more important was the other alternative which he had previously hinted at but now emphasized strongly. By a program of artificial insemination called "Voluntary Choice," Muller suggested that frozen sperm from genetically superior male lines was to be stored in sperm banks throughout the country. He pointed out that in couples where the man was impotent or sterile, artificial insemination had already become an established medical practice with complete social sanction. In the 1950's, over 10,000 such cases occurred in America every year. Why not, he argued, encourage these couples to use sperm from the genetically best males available rather than sperm of an average man. Each couple could choose the sperm line which they wished to use and would thus insure that their offspring would begin life with a sound genetic potential. Once the idea was established with these special cases, Muller felt it would be more easily extended to all individuals, giving every couple the opportunity to produce well-adapted and genetically fit offspring. As he wrote in 1961, "There is no physical, legal, or moral reason why the sources of the germ cells used should not represent the germinal capital of the most truly outstanding and eminently worthy personalities known" (Muller, 1961). In Muller's scheme, artificial insemination would lead to genetic therapy, which would direct human progress "toward the factors underlying creativity, initiative, originality . . . on the one hand and toward genuineness of human relations and affections on the other hand."

As time went on, and Muller became increasingly disturbed by human aggression, especially war, he modified his criteria for eugenic selection. In *Out of the Night* he had emphasized equally the traits of "comradeliness" and "intelligence." In his last public address, a paper presented at the International Congress on Human Genetics in Chicago in 1966, Muller stressed a selection of genes which favored cooperative behavior as being even more important to survival than intelligence (Muller, 1967).

The concept of voluntary choice was to Muller an important positive step toward insuring the spread of favorable genetic elements—whatever the collective wisdom of mankind decided they should be—to the population at large. In Muller's view, it was essential for the continued survival of man that he be convinced of the importance of providing the offspring of each generation with the best genetic foundation available. There was no value to the individual or to society, as he saw it, in the vain wish simply to pass on one's idiosyncracies to the next generation. As he had written in 1935: "Mankind has a right to the best genes obtainable . . ., and eventually our children will blame us for our dereliction if we have deliberately failed to take the necessary steps for providing them with the best if it was available, squandering their rightful heritage only to feed our heedless egotism" (Muller, 1935).

Without question, Muller's view of society was exalted, optimistic, and perhaps in many ways unattainable. His purpose was for society to direct itself toward the ends of improving its genetic makeup. While more technical information was needed—such as determination of the factors involved in inheritance of intelligence or cooperative behavior—this would come in time, just as information about the nature of variation and mutation had come. The most important and probably most difficult goal, as Muller saw it, was to convince society as a whole of the importance of eugenic aims.

"Voluntary choice" was Muller's specific program. It respected an individual's freedom, and was meant to be as its title implies, noncoercive. Through much lecturing and writing on this subject in the last decade of his life, Muller tried to insure that the seeds of his reforms would not fall on barren soil. He made a considerable impact, but his appeal was more to those who were already convinced, to the biologists, sociologists, and doctors, than to the large bulk of mankind, to whom his ideas ultimately most directly applied.

CONCLUSION

Through the interaction of his scientific and social ideas, H. J. Muller led the eugenics movement of the early 20th century away from its roots in conservative social philosophy. He accomplished this task first, by insisting on the necessity of sound biological knowledge for making any statements about regulation of human breeding; second, by emphasizing the voluntary and humanitarian aspects of eugenics; and third, by uniting ideas of genetic betterment with progressive social and medical views. The inevitable connection which Muller saw between heredity and environment made it imperative to set ideas of human evolution in a social and political context. The older social Darwinism and the early stages in the development of the eugenics movement lacked this balanced viewpoint. Guided by almost exclusively hereditarian attitudes, these ideas reflected the naivete of those who sought strict application of biological concepts to social organization. Muller's lasting contribution was to unite the hereditarian attitudes associated with traditional eugenics and the environmentalist's viewpoint associated with modern sociology to obtain a humane and reasoned approach to problems of controlled human breeding.

Muller lived at a crucial period in the development of ideas concerning eugenics and heredity. At the time he entered college (1906), the eugenics movement was at the height of its popularity; at the time he entered graduate school 4 years later, the new science of genetics was on the threshold of its first expansive development. Muller's own life-style in science not only preserved the influence of these two lines of thought but wove them together into a consistent social program which emphasized the necessary interplay between science and society. Thus, his early interest in eugenics gave direction to his later laboratory investigations in heredity. At the same time, his laboratory work in heredity and evolution spurred on and gave direction to his eugenic concepts.

Muller's career has shown clearly the complex interrelationships which existed in one man's mind between scientific and cultural ideas. More dramatically than most, Muller's represents what is a general phenomenon in the history of science: namely, that scientific ideas—often of the most technical sort—are not developed wholly in a cultural vacuum. They are influenced by and, in turn, influence social, political, and philosophical views.

While Muller's ideas may not have gained the widespread support he would initially have desired, he raised problems which mankind cannot escape facing. He continually approached the issues with the optimism of his youth that mankind could and someday would find meaningful answers. As he showed, the solution rested on the social and political—cultural, if you will—outlook which mankind would ultimately adopt. The solution which Muller himself proposed was based on scientific attitudes, that is, on the use of scientific knowledge to replace myth and egotism as bases for decisions about the future of human evolution. Muller realized that to accomplish this aim effectively a new cultural *milieu* would have to be created. The *milieu* which he envisioned—socialism and communism—would allow the most effective use of scientific concepts to better the condition of man himself. It was the necessity of the interplay of social and scientific views which guided Muller's thinking. This interplay, in historical perspective, provides an excellent example of the crucial interrelationship between biology and culture.

Notes

1. A recent study by William Morris. 1968. Hermann Joseph Muller: Radiation Genetics and Eugenics, A.B. Thesis, Harvard University, Cambridge, Mass., has been most helpful in my own thinking and as a partial guide to source materials on Muller's eugenic thought. Since the writing of this paper, a study of the early history of the eugenics movement in America has been published by Kenneth Ludmerer, 1969, American geneticists and the eugenics movement: 1905–1935. *J. Hist. Biol.,* **2:** 337-362. This article surveys clearly the initial enchantment of many geneticists with the Eugenics Movement, and their subsequent disillusionment with eugenic ideas and applications in the 1920's.
2. Extensive and valuable information about the development of the eugenics movement and its interrelation with social history can be found in the following works: Haller, Mark. 1963. *Eugenics: Hereditarian Attitudes in American Thought,* Rutgers University Press, New Brunswick, N.J.; and Hofstadter, Richard. 1955. *Social Darwinism in America,* The Beacon Press, Boston, Mass.
3. The progress of sterilization laws is summarized in Laughlin, H. H. 1926. *Eugenical Sterilization:* 1926. Yale University Press, New Haven, Conn. (Quoted from Hofstadter, p. 237.)

4. First published as *Die Continuitat des Keimplasmas als Grundlage einer Theorie der Vererbung* (Jena, Fisher, 1885). An English translation by Selmar Schonland was published as *The Continuity of the Germ Plasm as the Foundation of a Theory of Heredity* (Scribner's, New York, 1893). This translation is reprinted in August Weismann, 1891. *Essays upon Heredity and Kindred Biological Problems,* Vol. 1, E. B. Poulton, Selmar Schonland, and Arthur Shipley (eds.), Clarendon Press, Oxford. p. 167-254.
5. Lock's book was published in 1906 by E. T. Dutton Co., London. The influence of this book on Muller is recorded in Carlson's *The Legacy of H. J. Muller,* 1967. p. 438; also in a personal interview with Muller, by the author, March 1965, typescript p. 1.
6. From Muller's autobiographical notes, undated manuscript (probably 1941) quoted from Carlson, *The Legacy of Muller,* p. 440.
7. Morris, William. 1968. Hermann Joseph Muller: Radiation Genetics and Eugenics, A.B. Thesis, Harvard University, Cambridge, Mass.
8. H. J. Muller, "Revelations of Biology and their Significance," Unpublished address presented before the Peithologian Society, Columbia University, 1910. Manuscript in the Lilly Rare Books Library, Indiana University, referred to by Sonneborn, *H. J. Muller, Crusader for Human Betterment,* 1968, (p. 773.)
9. *Ibid.,* see esp. p. 773.
10. Peithologian Society paper, quoted from Sonneborn, T. M. 1968. H. J. Muller: tribute to a colleague. *The Review,* **11:** 19-23; esp. p. 20. Alumni Association of the College of Arts and Sciences-Graduate School, Indiana University (Fall 1968).
11. *Ibid.,* p. 20.
12. See correspondence on the matter, reprinted in Reingold, Nathan. Jacques Loeb, the scientist, his papers and his era. *Lib. Congr. Q. J. Curr. Acquis.,* **19:** 119-130. esp. p. 122-123.
13. Muller and Altenberg made this point explicit at the beginning of this article (p. 10). Interestingly, the paper is concerned solely with natural mutation rate, and not with artificially induced mutation.
14. It is beyond the scope of this paper to explain the details of Muller's exact procedure; these can be found in any standard genetic text-

book. For example, Srb, A. M., Ray B. Owen, and Robert S. Edgar. 1965. *General Genetics,* 2nd ed., W. H. Freeman & Co., San Francisco, p. 244-246; or, Muller's original paper, listed below.

References

Barker, Lewellys F. 1910. The importance of the eugenics movement and its relation to social hygiene. *J. Amer. Med. Assoc.,* **54:** 2018 ff.

Carlson, E. A. 1967. The legacy of Hermann Muller: 1890–1967. *Can. J. Genet. Cytol.,* p. 436-448.

______. 1968. Hermann Joseph Muller, a memorial tribute. *The Review,* **11:** 30.

Davenport, C. B. 1919. *Naval Officers, Their Heredity and Development,* Carnegie Institute of Washington, Washington, D.C.

Galton, Charles. (n.d.) *Inquiries into Human Faculty and its Development.* 2nd ed., E. P. Dutton & Co., New York, p. 17, fn.

Haller, Mark. 1963. *Eugenics: Hereditarian Attitudes in American Thought,* Rutgers University Press, New Brunswick, N.J.

Hofstadter, Richard. 1955. *Social Darwinism in America.* The Beacon Press, Boston, Mass.

Jordan, D. S. 1911. *The Heredity of Richard Roe,* American Unitarian Assoc., Boston, Mass., p. 35. Referred to in Hofstadter, 1955, p. 164.

Ludmerer, Kenneth. 1969. American geneticists and the eugenics movement: 1905–1935. *J. Hist. Biol.,* **2:** 337-362.

Morgan, T. H., A. H. Sturtevant, H. J. Muller, and C. B. Bridges. 1915. *The Mechanism of Mendelian Heredity.* Henry Holt, New York.

Muller, H. J. 1927. The effects of X-radiation on genes and chromosomes. *Science,* **67**: 82-85.

______. 1929. The method of evolution. *Sci. Monthly,* **29:** 401-505.

______. 1934. Radiation genetics. *Proceedings of the 4th International Radiologen Kongress 2,* p. 100-102.

______. 1935. *Out of the Night: A Biologist's View of the Future.* Vanguard Press, New York.

______. 1961. Human evolution by voluntary choice of germ plasm. *Science,* **134:** 643-649.

———. 1962. Radiation genetics, *Studies in Genetics, the selected papers of H. J. Muller* (reprinted from *Proceedings of the 4th International Radiologen Kongress 2* [1934]). Indiana University Press, p. 29-293.

———. 1967. What genetic course will man steer? *Proceedings of the Third International Congress on Human Genetics 1966,* The Johns Hopkins University Press, Baltimore, Md., p. 521-543.

Muller, H. J., and E. Altenberg. 1919. Rate of change of hereditary factors in *Drosophila. Proc. Soc. Exp. Biol. Med.,* **17:** 10-14.

Muller, H. J., and J. T. Patterson. 1930. Are progressive mutations produced by X-rays? *Genetics,* **15:** 495-578.

Sonneborn, R. M. 1968. H. J. Muller, crusader for human betterment. *Science,* **162:** 772-776.

B.
Basic Genetic Patterns

This section contains papers which explore relationships: individual, familial, and social through the simpler genetic patterns of dominance, recessiveness, and twin development. The reports indicate that social implications are derived from genetic patterns and genetic implications from social contacts.

In recent years, researchers have developed serological methods whereby human remains can be "blood typed." In their paper Harrison, Connolly, and Abdalla use these methods on two mummies to confirm that there is a familial relationship between them. The researchers test the remains of the presumed brothers for ABH blood group substances and MN substances.

In principle the research plan in the paper is based on some simple propositions. The greater the similarity of blood group results, the greater is the likelihood that the two individuals are related. Conversely, the greater the blood groups are dissimilar, the greater is the likelihood that the two individuals are not related. Specifically, the more blood group systems an investigator can use for determining similarities and dissimilarities, the greater is the likelihood that he/she could establish whether individuals come from the same family. While more detailed group methodologies are used to settle disputed paternity suits involving presumed parents and offspring, the basic principles are the same. Given adequate information the geneticist can determine the probability of presumed relationships.

Many researchers use identical twins in human twin studies to differentiate between hereditary and social contributions to human development. In working with identical twins, the most effective way of establishing monozygosity is to use blood grouping methods similar to those found in the paper by Harrison and his colleagues. An investigator undertakes many blood group tests involving the ABO, MN, Rh systems, etc., then establishes the exact blood group type of each individual in each of these systems. If many test results show two individuals have identical data, then it can be said with reasonable certainty that the two people were derived from the same fertilized egg. In addition other factors such as anatomical characteristics of the two individuals and, if available, the condition of placental mechanism at birth are utilized in twin determination.

The paper by Abe shows how researchers obtain data on personal characteristics of twins. Abe is correct in saying that his research represents only an exploratory investigation. However, it is a good initial study which supplies leads for further research. The paper is instructive for students because it shows the occasional difficulty one encounters in obtaining concise and definite data from twin studies.

Keeler has studied the Cuna Indian population of Panama for approximately 20 years. His fine studies have resulted in numerous papers and it was most difficult to choose only one for inclusion in this volume. The reader is also encouraged to read the paper by Keeler listed in Additional Readings. Keeler has developed an important sociogenetic analysis of a group of albinos and their interactions with their society.

Anthropologists tell us that many cultures have folktales concerning the origin of albinism. The Cuna Indians call them moon-children because it was believed that the mother or father spent an inordinate amount of time looking at the moon during the child's gestational period. We know, of course, that albinism is generally due to a single pair of recessive mendelian mutant genes. Over a number of years of investigation, Keeler appears to be satisfied that this basic pattern accounts for moon-child albinism.

In a short paper by Turner, there is a remarkable amount of sociogenetic interpretation. Using the familiar technique for PTC tasters and nontasters, Turner finds the PTC distribution in fraternity and sorority members quite different than one taken from a random student population. In simplistic form we have:

Social institution is formed which
↘ attracts individuals with similar genetic patterns and which also
↘ establishes potential mating situations whereby there is
↘ marriage and subsequent perpetuation of the distribution of genetic traits
↘ originally found in the social institution.

The above situation is hardly unique. Groups of young people of marriageable age are likely to come together for a variety of social reasons. If one ethnic group, for example, develops a social organization, there is a great likelihood that marriage and subsequent replication of the gene pool of that ethnic group will take place. An example of these on college campuses today are Afro-American Centers and Hillel Foundation Centers.

Additional Readings

Connolly, R. C., 1969. Microdetermination of blood group substances in ancient human tissues. *Nature,* **224**:325. This paper has further details about the methodology used in paper No. 2.

Keeler, C., 1970. Cuna moon-child albinism, 1950-1970. *Journal of Heredity,* **61**:273-278.

Boon, R. A. and D. F. Roberts, 1970. The social impact of haemophilia. *Journal of Biosocial Science,* **2**:237-264. The authors discuss how haemophilia, one of the oldest known genetic disorders, presents certain social limitations to those who are afflicted with the disease. Regrettably, this fine paper is too long for inclusion in this volume.

Becker, Kenneth L., 1967. A twin study method in medicine and genetics. *Postgraduate Medicine,* June, 603-609.

Hurst, L. A., 1970. The twin-family method in psychiatric genetics illustrated from the investigations of Franz J. Kallman. *Acta Get. Med. Gemellol,* **19**:135-139.

Abe, L., 1969. The morbidity rate and environmental influence in monozygotic co-twins of schizophrenics. *British Journal of Psychiatry,* **115**:519-531. Further readings on the twin study method and how it contributes to our knowledge of heredity and the environment.

Questions

If Smenkhkare and Tutankhamen were brothers, what were the parental possibilities for blood typing in the ABH and MN systems?

What social and genetic factors could increase the incidence of albinism? Decrease the incidence? How does our own culture view albinos?

Referring to Turner's study, which social and cultural organizations would tend to perpetuate specific gene distributions? Which would tend to disperse and bring in more genetic variations to the gene distributions?

Kinship of Smenkhkare and Tutankhamen Demonstrated Serologically

R. G. HARRISON, R. C. CONNOLLY, AND A. ABDALLA

Recent researches[1] have demonstrated that the human remains in the Museum of Antiquities in Cairo, formerly thought to belong to Akhenaten (Amenophis IV), are more likely to be identified with Smenkhkare, co-regent and successor of Akhenaten as pharaoh of Egypt in the XVIIIth dynasty.

Because of the suggestions that Smenkhkare and Tutankhamen were brothers[2], and because the remains of Smenkhkare show several interesting anthropological features, it was important to re-appraise the anatomy of Tutankhamen, and this was done in December 1968. Detailed findings will be published elsewhere and it is sufficient here to note that there are very many points of similarity. Indeed, certain anthropometric measurements are identical. In order to investigate further the degree of kinship between these two pharaohs, it was clearly essential to estimate their blood groups.

It is well known that the ABH blood group substances occur not only on the red blood corpuscles but are also widely distributed throughout the tissues of the living body. The polysaccharide nature of these substances renders them highly resistant to climate and microbial degradation and they can be readily demonstrated in even very ancient human remains. It appears that the same can be said for the distribution and persistence of the MN substances but probably not for the antigens of the Rhesus system.

The sensitive technique developed in this department and described above has revealed blood group substances of the ABH and MN systems in tissue from Tutankhamen, Smenkhkare and other mummified remains. At present, it has not been possible to demonstrate with certainty any Rhesus antigens from mummified material.

The results of these investigations suggest that both Tutankhamen and Smenkhkare are blood group A_2 and that they are both MN. It is considered that these results are specific to the blood groups of the individuals concerned and not to microbial contamination for instance.

Reprinted by permission from *Nature,* 1969, Volume 224, 325-326.

Interpretation of these findings is difficult in three respects. First, the A_2 group could have been A_2B and the B substance has degenerated, but this is not very likely. The genotype could have been A_2O or A_2A_2 or A_2A_1. This is unlikely because, as with A_2B, there is no reason to suppose that the A_2 substance is more persistent than A_1 or B; with this material, however, it is not possible to distinguish between A_2O and A_2A_2 genotypes. The final difficulty relates to the incidence of these blood groups in Egypt more than 3,000 years ago. Although blood group distributions remain remarkably constant for long periods of time, the present day distributions in Egypt cannot be taken as representative of the dynastic situation. In other words, although both sets of material suggest the same blood groups, these may have been the only blood groups in Egypt at the time. Other workers have, however, examined material of a similar origin and have demonstrated other ABH blood groups, which adds significance to the present results.

These palaeogenetic findings alone do not in themselves prove any relationship between Tutankhamen and Smenkhkaro but do increase the probabilities. Further studies with this and other similar material are in progress which it is hoped may throw more light on this interesting problem.

We thank the Department of Antiquities, Cairo, and in particular Dr Gamal Moktar, Dr Gamal Mehrz, Dr Z. Iskander and Dr Henry Riad for their help with this investigation. We also thank the Liverpool Regional Blood Transfusion Service for facilities.

References

1. Harrison, R. G., *J. Egypt. Archacol.,* 52, 95 (1966).

2. Desroches-Noblecourt, C., *Tutankhamen* (Michael Joseph Ltd, London, 1963).

3. Reactions to Coffee and Alcohol in Monozygotic Twins*†

K. ABE

"... legend credits the discovery of coffee to a friar of an Arabian convent. Shepherds reported that goats which had eaten the berries of the coffee plant gamboled and frisked about all through the night instead of sleeping. The friar, mindful of the long nights of prayer which he had to endure, instructed the shepherds to pick the berries so that he might make a beverage from them. The success of his experiment is obvious from the fierce opposition stirred up by the more orthodox section of the priests against the use of this devotional antisoporific, and from the popularity of coffee today. A billion kg are consumed annually in the United States alone.

Modern pharmacological studies of caffeine have amply confirmed the ancient belief that caffeine has a stimulant action, which is the basis for the popularity of all the caffeine-containing beverages."

Goodman and Gillman [1]

During the course of an investigation of 11 pairs of monozygotic twins, one or both of whom had suffered from an affective disorder, the opportunity was taken to record the reactions of the twins to social doses of coffee and alcohol as reported by the twins and their relatives. Although these data are rather scanty, it is thought they may be of general interest as no previous studies relating to the inheritance of the behavioural response to alcohol and coffee have been reported. This neglect of the hereditary aspects of responses to such popular and widespread drugs may seem surprising. But it may also be said that even straightforward descriptive reports on the behavioural effects of alcohol and coffee, and the individual variation in these reactions in our population, are very scanty. This neglect has recently been pointed out by Elkes [2].

In the case of coffee, marked variations in behavioural response have been reported Goldstein *et al.* [3,4]. Compared with placebo decaffeinated coffee, decaffeinated coffee containing 300 mg of caffeine

*From the Maudsley Hospital, Denmark Hill, London S.E.5.
†Present address: Osaka City University Medical School, Asahi-machi, Abeno-ku, Osaka, Japan.

Reprinted by permission from *Journal of Psychosomatic Research,* 1958, Volume 12, 199-203.

increased the time taken to get to sleep in some subjects and tended to increase alertness and physical activity. In other subjects there was an increase in nervousness, and in still others the drug was without any effect at all. The variation in these effects was not due to variation in plasma levels or rate of metabolism or excretion of the behavioural effect.

Controlled descriptive studies of the effects of alcohol in man are even more scanty, perhaps because the effects and their variability are so well known to everybody. There are thought to be considerable variations in the changes in skin colour, sleepiness, mood, aggression, and, above all, in sociability. The response may, of course, vary with the circumstances in which the drug is taken, but Takala *et al.* [5] are of the opinion that there is a characteristic behavioural response for each individual. If each individual has a characteristic response then it becomes important to ask to what extent this characteristic is determined by hereditary factors.

Since alcohol and coffee affect mood and behaviour, they are of interest to psychiatry. Moreover, there may be a more specific relation to mental illness. Some patients who have recovered from acute schizophrenia or mania may, after drinking alcohol, show a transitory behavioural disturbance in some way reminiscent of the clinical picture they manifested during the illness. Disturbances of sleep are a feature of many mental illnesses, and the interesting possibility arises that the tendency to manifest insomnia in mental illness may be related to the tendency to manifest insomnia after drinking coffee; or even that insomnia after coffee may be a clue to some slight increased disposition to mental illness.

Method

The subjects were drawn from a register of all twins treated since 1948 at the Bethlem Royal and Maudsley Hospitals, whether as inpatients or outpatients. Only monozygotic pairs were selected, in which one or both twins had been diagnosed as suffering from affective disorder and it was considered that both twins might be suitable and available for investigations of electrolyte balance [6]. Monozygosity was established by methods which included full blood-grouping. For the purpose of the present paper 11 pairs were available. Six of the pairs were male, and the ages ranged from 27 to 65 at the time of investigation.

Each twin was questioned about his responses to alcohol and coffee, with particular emphasis on changes in activity, mood, sleep, and autonomic reactions. Each twin was also asked about the reactions of his

Table 1. Clinical features of the twins, and reactions to alcohol and coffee

		Illness	Reaction to coffee		Reaction to alcohol		
(1) Age	(2) Sex	(3) D = depression M = mania P = paranoid – = never ill	(4) + = onset w. insomnia ++ = marked insomnia	(5) + = insomnia – = sleep not affected	(6) r = markedly red in face – = not	(7) + = over-talkative – = not	(8) h = happy and placid – = not happy l = emotional lability
48	m	MP	++	+	–	+	–
		DM	++	+	–	+	–
62	m	DP	++	+	–	–?	–
		–		+	–		h
65	m	DP	–	+	–	–	–
		MP	–	+	r	–	–
42	f	D	++	+	Both abstainers		
		D	++	+			
46	f	–		+	r	–	h
		DP	–	+		–	–
44	f	D	–	–	r	–	–
		DP	–	–	r	–	–
27	m	D	+	–	r	+	–
		–		–	r	–	–
37	m	DP	+	–	r	+	h
		D	+	–	r	+	h
53	f	MD	+	–	r	+	hl
		–		–	r	+	h
57	f	D	+	+	r	+	hl
		–		–	r	+	h
65	m	D	+	–(+)	r	–	h
		–		–	r	–	h
Significance of intra-pair concordance		N.S.	(5 pairs) 0.038	0.0059(0.035)	0.10	0.012	0.055

co-twin and additional information was obtained from husbands and wives. The information provided by the different informants tended to be in substantial agreement.

The clinical features of the affective illnesses were already well documented and were abstracted from the casenotes.

Results

The main results are shown in Table 1. In the case of coffee, the apparently most reliable information concerned the question of whether taking coffee before bedtime disturbed the subsequent sleep. This appeared to be an either/or effect (column 3). It will be seen that the pairs are distributed fairly evenly between 'sleepless' and 'no effect' and only one pair is discordant. In another pair only one twin became mentally ill and this twin alone developed sleeplessness after coffee, but only after his illness. This intrapair concordance in reaction to coffee is significant at the 1 per cent level (Fisher's exact test).

Since one pair of twins were both abstainers, there were only 10 pairs to study for alcohol reactions and these were less significant than in the case of coffee. Concordance for overtalkativeness (column 5) was significant at the 5 per cent level. Here again the pairs are fairly evenly distributed between over-talkative and not over-talkative; one pair was discordant and one doubtful. There was a tendency towards concordance for the reaction of happiness as opposed to no reaction or a nasty suspicious and jealous reaction to alcohol, but this was not significant. Since most of the subjects reported the development of a red face after alcohol, the data do not give a fair test of the significance of the intrapair concordance of this trait.

The main clinical features of the affective illnesses are shown in column 3 of Table 1. There was no apparent association between any feature and response to the drugs.

In column 2 is shown the presence of insomnia at the onset of or preceding the affective illness. This refers to insomnia over and above any insomnia normally experienced by the subject. Of the 5 pairs who had both been ill, all were concordant for this trait; in 3 pairs both manifested insomnia, in 2 pairs neither. The concordance is significant at the 5 per cent level.

There was no association between sleeplessness after coffee and the manifestation of insomnia at the onset of the illness. However, there were one or two associations between the variables (Table 2). The patients with paranoid features in their illnesses tended either to become jealous and suspicious after drinking or at least not to show the common happy reaction; alcohol did not make them red in the face. Those who did become red in the face after alcohol tended not to have insomnia after coffee. Those whose illness started with insomnia tended to become over-talkative after alcohol. Of course there are many pairings possible between these variables and some of these associations are no doubt fortuitous.

Table 2. Significance of associations between the variables (Fisher's exact test)

	Paranoid features	Insomnia at onset of illness	Insomnia after coffee	Red face after alcohol	Talkative after alcohol
Not happy after alcohol	0.0054	N.S.	N.S.	N.S.	N.S.
Talkative after alcohol	N.S.	0.0030	N.S.	N.S.	
Not red after alcohol	0.0029	N.S.	0.0022		
Insomnia after coffee	N.S.	N.S.			
Illness started with insomnia	N.S.				

Discussion

The present investigation is exploratory in nature and the results must be viewed with caution. The numbers are small and all the pairs were monozygotic. In order to separate the effects of heredity from environment the normal procedure is to compare the concordance rates in monozygotic twins with the concordance rates in dizygotic twins. We have not studied dizygotic twins, and therefore the concordance of the monozygotic twins could possibly be due to some common environmental factor. However, many of the twins are over 40 years of age and have been living apart for about 20 years. One might have expected drug reactions determined by common enviromental effects to wane after this period of separation; and it must also be borne in mind that there were relatively slight changes in reaction induced by the affective illnesses.

Again, because of the nature of the investigation, the interviewing of the twins was not as systematic as might have been desirable. Ideally, a different investigator should examine each twin and make a judgement of the drug reactions independent of any knowledge of the reactions of the co-twin. In the present circumstances this was not possible. Subject to the above reservations therefore, we may suggest that there is a significant genetic variation in sleep disturbance after drinking coffee, over-talkativeness after drinking alcohol and the presence of marked sleep disturbance in the initial phase of an affective illness. The relation of these drug reactions to other drug reactions and to other components of the behavioural repertory of the individual remain to be explored. However, one interesting association might perhaps be mentioned. Abe and Shimakawa [7] reported that those adults who reported difficulty in sleeping after coffee had tended as children to have difficulty in falling asleep or to sleep lightly even without coffee. Thus it is possible that this hypersensitivity to caffeine is one manifestation of a fairly unspecific low threshold of the central nervous system to excitation.

None of these reactions differentiated the ill twins from the well twins, but the number of discordant pairs is so small that an association would have had to be extremely marked to appear in these data. There is an interesting suggestion, however, that from the response to alcohol one might predict the type of psychosis which would occur in that there was some tendency for those who became suspicious and jealous after alcohol to have developed paranoid symptoms in their affective illness, whereas those who became placid and happy after alcohol tended to be free from paranoid features. But the difference is not great and is only significant when the never ill twins are included in the analysis.

Summary

In an exploratory investigation, 11 pairs of monozygotic twins were asked about their reactions to alcohol and coffee. There was a significant within-pair concordance for insomnia after drinking coffee and over-talkativeness after drinking alcohol. One or both of each pair had suffered an affective illness; there was an association between over-talkativeness after alcohol and insomnia at the onset of the illness (the latter trait also showed a significant within-pair concordance). Those whose skin reddened markedly after alcohol tended to sleep well after coffee.

Acknowledgments

The author is grateful to Dr. E. Slater, Mr. J. Shields and my colleague Dr. J. S. Price for valuable suggestions and advice and to Miss V. G. Seal for her assistance in follow-up study.

References

1. Goodman L. S. and Gillman A. (Eds.) *Pharmacological Basis of Therapeutics,* Macmillan, New York (1965).

2. Elkes J. Behavioral pharmacology in relation to psychiatry. *Psychiatrie der Gegenwart.* Vol. 1. Springer, Berlin (1967).

3. Goldstein A., Warren R. and Kaizer S. Psychotropic effects of caffeine in man. Part 1. Individual differences in sensitivity to caffeine-induced wakefulness. *J. Pharmac. Exp. Ther.* **149,** 156 (1965).

4. Goldstein A., Warren R. and Kaizer S. Psychotropic effects of caffeine in man. Part 2. Alertness; psychomotor coordination and mood. *J. Pharmac. Exp. Ther.* **150,** 146 (1965).

5. Takala M., Pihkahen T. A. and Markhen T. *The Effect of Distilled and Brewed Beverages,* The Finnish Foundation for Alcohol Studies, Helsinki (1957).

6. Coppen A. J. and Abe K. Paper in preparation.

7. Abe K. and Shimakawa M. Genetic-constitutional factor and childhood insomnia. *Psychiat. Neurol.* **152,** 363 (1966).

4. The Incidence of Cuna Moon-Child Albinos

CLYDE KEELER

Several authors have attempted to count the albinos in the Caribe Cuna Indian population of San Blas Province, Panama[1-5]. Each has been amazed at their high incidence, but little has been offered by way of explanation for such a concentration of albinism. (The Cuna call them "Moon-children" and believe they are caused by either the mother or father looking at the moon too much during gestation.) Probably no account is highly accurate, because of certain difficulties that will be discussed in this paper, but estimates can be sufficiently precise to warrant a few conclusions.

Such counts have usually been made by visiting the larger, more civilized islands, recording the albinos there, and then, through inquiry, learning of the presence of albinos on other islands. I have visited the coast that includes the inhabited islands from Mandinga to Titumate. Reports on the few hostile mountain towns had to be obtained from traveling residents of those towns.

Wafer, the pirate, (ca 1681) estimated that the incidence of albinism among the Cunas was 0.3 to 0.5 percent, or 30 to 50 per 10,000 population[5] (Table 1). This estimate has been followed by the work of Harris[1], Stout[4] and Keeler[2].

Table 1. Estimates of Cuna Indian populations and albinism

Author	Date	Total population	No. of albinos	Albinos per 10,000 population
Wafer	1681	?	?	30-50
Harris	1925	20,100	138	69
Stout	1940	20,831	98	47
Keeler	1950	22,822	152	67
Keeler	1962	23,743	144	61

Reprinted by permission from *Journal of Heredity,* 1964, Volume 55, 115-120.

Fig. 1 Cuna albino schoolboys.

It will be noted in Table 1 that Stout[4] recorded fewer albinos than did Harris or Keeler. Stout accounts for this by assuming that his own figures are more accurate than those of his predecessor, Harris. However, from the consistency of Harris's 1925 studies with mine of 1950 and 1962 —where the incidences are 69/10,000, 67/10,000 and 61/10,000 respectively—it would appear that Stout was probably the one in error and Harris's estimates were fairly accurate.

INFANTICIDE

The practice of infanticide complicates the problem. Supreme Chief Nele Kantule, a prestigous leader of the Cunas, actively and constantly preached to the Cunas during his rule that because the albinos sin less than other Indians, they are on better terms with Tiolele, the Sungod, and for this reason they should not be killed at birth, but should be kept alive so that they might tell the Cunas more about God.

Before the time of Nele Kantule the percentage of albinos kept alive was probably much lower than during his rule, as suggested by the incidence estimated by Wafer.

Following his death, however, the incidence of albinos seems to have declined outside of the large and more civilized towns, and various Indians tell me that there is now a considerable amount of infanticide, especially on the smaller, less civilized islands. This recently increased practice of infanticide would help to account for the drop from 67 per 10,000 in 1950 to 61 per 10,000 in 1962.

Although we expect to find some variation in the albino gene distribution on different islands, we may check our assumption of infanticide by comparing the incidence of albinism in 1962 for more civilized Cuna towns with that of less civilized towns.

The data from the more highly civilized towns—Ustuppo, Ailigandi, Mulatuppo, Nargana, Corazon de Jesus (Nusatup) and Ignacio de Tupile—where the killing of newborn children is frowned upon, show a high rate of albinism (Table 2A). Among these towns a population of 6,365 persons includes 92 albinos, or 144 per 10,000. Nargana, the town that has been transformed most by changing acculturation, shows the astounding rate of 475 albinos per 10,000 population.

We note that the albino incidence is not as high in a group of the larger towns that are considered to be less civilized than those in (Table 2B). Among the 2,594 persons in less civilized towns we find 14 albinos, or 54 per 10,000.

Thus, there is more than twice the incidence of albinos in the more civilized large towns than in the smaller, less civilized ones.

Figures for the uncivilized small towns and islands where only a few families were present, where infanticide could be hidden easily, and where there was little or no social pressure against this time-honored custom, show an even more striking reduction in the incidence of albinism (Table 2C). The 68 other Cuna localities, ranging from 3 to 746 inhabitants, as recorded in the 1960 census, have 33 albinos in a total population of 10,411, or 32 per 10,000. Thus, the statements of various

Fig. 2 Cuna albino school child. She has completed a fifth grade Mission School education and is employed as a secretary in the Clinica Marvel Hospital at Aligandi.

Fig. 3 Moon-child brothers and their brown sister.

Indians that infanticide is practiced on newborn albinos, especially in the small villages, is substantiated by the data collected.

Table 2. Incidence of albinism in towns selected as civilized, less civilized and uncivilized

Name of town	Number of Moon-children	Population	Incidence per 10,000
A. Civilized:			
Ustuppo	8	1,705	40
Mulatuppo	14	1,383	101
Ailigandi	24	1,355	177
Tupile	8	868	92
Nargana	30	632	475
Corazon de Jesus	8	423	189
Total	92	6,365	144
B. Less civilized:			
Plyon Chico (Okup Seni)	5	1,178	42
Achutuppu	6	781	76
Mammituppu	3	635	47
Total	14	2,594	54
C. Uncivilized:			
68 localities ranging in size of population from 3–746	33	10,411	32

The small, uncivilized localities have more than four times fewer albinos than the large, more civilized towns, and almost 60 percent fewer than in the quasi-civilized large towns.

It will be noted that the figure 32/10,000 is actually near the lower limit of Wafer's estimate of 30 to 50 per 10,000 for the uncivilized Cunas of his day.

ABSENTEES AND CENSUS RECORDS

Aside from the practice of infanticide, the method of recording the censuses taken by the Panamanian Government is a second factor that distorts the observed incidence of albinism among the Cunas. Official census records do not include as Cunas any members of the tribe who may have found employment outside the Reservation. They are recorded as Panamanians.

In 1925, Harris[1] had to estimate the population of San Blas, as there was no census report available to him at that time. In 1940, Stout[4] added 4,000 working absentees from the Reservation to the 1940 census population record for San Blas.

In 1950, in order to adjust the population figures of the census, Dr. Alcibiades Iglesias (an educated Cuna) and I estimated, on the basis of the number of Cunas seen on the streets of Colon and the City of Panama, that there were then 2,000 absentees from the Reservation. In 1960, Iglesias and I estimated that the number of absentees had increased to 4,000. However, the number of absentees varies from time to time.

For 1962, I have taken the 1960 census figure, added 4,000 for absentees, and an additional 400 for normal population increase.

It is felt that Stout's estimate[4] of 4,000 absentees may have been too high, since his number of albinos is definitely too low when compared with the figures of Harris[1] and Keeler[2].

Stout compared his 1940 data on the percentage of albinos for several towns with the 1925 data of Harris. We continue this comparison for 1950 and 1962 in Table 3.

All towns except Nargana seem to hold their incidence of albinism at a constant number. Albinism in Nargana, however, has steadily increased, although its population is now about one-half of what it was in 1925.

VIABILITY

The old men say that albinos, on the average, do not live as long as normally-pigmented Cunas[1]. Indeed, Chief Lacenta told this to Pirate Wafer almost 300 years ago. The validity of this belief may be examined statistically. Out of 156 normal Indians whose ages were recorded, 51 or 32.7 percent survived to 40 years of age, according to census data. Out of 74 albinos whose ages were recorded, four or 5.4 percent survived to 40 years of age. Taking the 32.7 percent as normal, the albinos have a survival value of 16.5 percent of normal at age 40. The breakdown shows that 117 normal females had a mean age of 22.5$\pm$ 1.5 years and 197 normal males had a mean age of 21$\pm$ 1.1 years. From our 1962 data we learn that 26 albino females have a mean age of 18.0$\pm$ 1.6 years, and 48 albino males have a mean age of 16.3$\pm$ 1.6 years.

Table 3. Incidence of albinism in specific towns over the years

	Harris 1925			Stout 1940			Keeler 1950			Keeler 1962		
	No. of albinos	Pop.	Albinos per 10,000	No. of albinos	Pop.	Albinos per 10,000	No. of albinos	Pop.	Albinos per 10,000	No. of albinos	Pop.	Albinos per 10,000
Ailigandi*	12	1250	96	14	605	231	21	1218	172	24	1354	177
Ustupo	5	1000	50	4	1139	35	8	1356	59	8	1705	47
Nargana	8	1200	67	12	1029	166	24	608	394	30	632	475
Tupile	2	250	80	4	622	64	10	752	133	8	868	92
Plyon Chico	2	300	67	2	576	26	2	902	22	5	1178	42
Tigre	2	150	133	4	573	70	3	550	54	4	671	60

*Population figures for these towns were given upon request by the Panamanian Census Bureau and differ somewhat from their published figures.

Table 4. Calculations from the Cuna pedigree data, by year

Date	Total pedigree population	Albinos calculated	Albinos found	χ^2	Probability
1950	269	95.291	103	0.694	.30–.50
1960	286	103.581	96	0.870	.30
1962	269	102.327	111	1.186	.05–.30

Several factors are involved in the Cuna sex-specific death rates. According to our records, females enjoy a more protected life than males, thus achieving a greater age. On the other hand, it is estimated that one-third of Cuna girls, on the average, will die in childbirth, a factor tending to reduce the mean age of the girls.

It will be seen that the mean age for albino females is 18 years compared with 22.5 years for normal females, although the difference may not be statistically significant.

Albinos seldom marry, although a few albino girls are married and others have illegitimate children. The rareness of childbearing extends the mean age of those albino girls that reach a reproductive age. All albino girls lead sheltered lives, compared with normal girls.

The normal boys are subject to environment hazards, such as poisonous plants and animals, and occupational accidents. The albino boys are somewhat protected, and their reduced mean age is evidence of their reduced viability.

GENE FREQUENCY

The data collected in 1950 and 1962 show a decline of the actual number of albinos during the past 12 years. The frequency has declined from 67/10,000 to 61/10,000. (Chief Nele Kantule, who preached against infanticide, died in 1944.) However, considering the results of complete selection at the end of one generation, with the albinos excluded from the mating pool, and without mutation, we may calculate from the formula

$$q_1 = q_0/(1 + q_0), \qquad (1)$$

an expected frequency of 57/10,000 for the succeeding generation instead of the 61/10,000 that we found. Assuming 18 years between generations (since the average female is dead at 22) the incidence in 1940 should have been over 67 and that in 1925 should have been over 82.5; or, calculating in the other direction, we have too many albinos today, according to the calculated mutation rate. Nevertheless, it is unsafe to press these conclusions too far because of the necessity of estimating absentees, although we feel relatively certain that nearly all of the albinos have been counted.

MUTATION RATE

By use of the formula $M = f(1 - F)$ it may be calculated that something in the neighborhood of 398 mutations to Moon-child albinism are added to the gene pool per 10,000 each generation of about 18 years duration. This would result in about five new albino homozygotes per generation in order to keep the incidence at equilibrium.

However, according to the formula (1), in the two generations between 1925 and 1962 there should be a greater reduction in the incidence of albinism than was observed.

While we may have indications and hunches from the mathematical treatment of our data, we dare not draw absolute conclusions because of the estimated approximations that cannot be determined precisely such as: 1) number of absentee workers, 2) number of siblings, 3) number of infanticides, 4) number of years between generations, and 5) exact ages of older persons for whom the date of birth was not recorded.

THE 3:1 RATIO

My 1950 calculations[2] from the pedigree date, using the formula

$$\alpha\alpha = .25/[1 - (.75)^n]$$

to adjust for the Hardy-Weinberg law, may be compared with my 1960 and 1962 data and calculations (Table 4). It will be noted that in 1960 there were fewer albinos than calculated but that in 1950 and 1962 there were more than expected. However, in spite of the tendency to eliminate some albinos today, we have more albinos in the pedigree data than expected on the basis of a 3:1 ratio. This may be partly due to the psychological tendency of informants to remember the affected persons in a sibship but to forget the normal siblings.

SUMMARY

The incidence of Cuna Indian Moon-child albinos is more than 60 per 10,000 population, being reduced from about 69 per 10,000 in 1925. Stout's 1940 estimate of 47 per 10,000 is probably in serious error. Our 1962 data suggest infanticide in uncivilized towns and possibly selective interbreeding of heterozygotes in the town of Nargana. Absolute population figures cannot be obtained, due to census methods and unrecorded absentees. Our figures show albinos to be less viable than normals, in keeping with general Cuna belief. Since albinos seldom reproduce, the gene frequency in one generation without mutation should have been reduced to 57 per 10,000, so there is evidence that mutation does take place. It is estimated that about 398 mutations per generation per 10,000 population are added to the gene pool in order to provide the five albino homozygotes per generation necessary to keep the incidence at equilibrium. Examination of the 1950, 1960 and 1962 data all show Moon-child albinism to be due to a single pair of recessive, Mendelian, mutant genes.

Literature Cited

1. Harris, R. G. The San Blas Indians. *Am. Jour. Phys. Anthrop.* 9:1-15. 1926.

2. Keeler, C. E. The Caribe Cuna Moon-child and its heredity. *Jour. Hered.* 44:163-171. 1953.

3. McFadden, A. W. Skin disease in the Cuna Indians. *Arch. Derm.* 84:175-185. 1961.

4. Stout, D. B. Further notes on albinism among the San Blas Cuna, Panama. *Am. Jour. Phys. Anthrop.* n. s. 5:483-490. 1941.

5. Wafer, L. A New Voyage and Description of the Isthmus of America. 1681. Reprint. Cleveland. Burrows Bros. 1903.

5. PTC Tasting in Two Social Groups

CHRISTY G. TURNER II

The purpose of this brief note is to add support to the view that cultural factors, such as racial and social discrimination, can affect the genetic composition of human groups by selective membership criteria.

A sample of University of California introductory physical anthropology students was tested some time ago on a volunteer basis for their ability to taste PTC (phenylthiocarbamide) by the 5% paper-impregnation technique. Students in these large introductory classes are from nearly every department, and the sample can be viewed as effectively random for this student population. All geographical and many local races are represented on campus and in these courses, although the majority in both instances are Caucasian (white). At the same time the above sample was obtained, members of a Greek-letter social fraternity and a sorority were tested by the same method. This group is largely white American.

Table 1

Group	Non-taster (tt)	Taster (TT,Tt)	Total
Random students	24.2%	75.8%	161
Fraternity and sorority members	37.8	62.2	82

The results of the two sets of tests are shown in Table 1. What is obvious by simple inspection is that the two samples (random and fraternity) are different in their frequency of non-taster/taster phenotype ratios. A chi-square test shows that the difference is statistically significant ($x_1^2 = 4.88$; P is between 0.05 and 0.01). There is little difference between the male fraternity non-taster frequency (38.7%) and that of the female sorority (37.2%), so the difference in Table 1 cannot be attributed solely to sex influence on the PTC polymorphism.

Reprinted by permission from *Social Biology,* 1970, Volume 17, No. 2, 142.

The fraternity sample is significantly different from that of the random sample largely because of *de facto* and/or overt racial discrimination, which at the time these data were collected was a major campus issue. The PTC nontaster phenotype frequency of the fraternity group is more like that of Europeans (*c.* 25 to 45%) and American whites (*c.* 30 to 35%) than like that of American Negroes (9 to 23%), African Negroes (*c.* 6%), Mongoloids (7 to 10%), or American Indians (<3%) according to Montagu (1960, p. 222). Further, these PTC data suggest that the genetic effects of social selection may be more pervasive than commonly supposed, making the evaluation of natural selection on human variation all the more difficult.

Acknowledgments

J. Cadien, M. Winton, K. Gibson, L. Fritz, and M. MacRoberts are gratefully acknowledged for their help.

References

Montagu, A. 1960. *Human Heredity.* Mentor. New York.

C.
Human Chromosomes and Antisocial Behavior

Recently, scientific and medical journals as well as the popular press have printed many articles on chromosomes and behavior. It is an extremely important area to understand and some basic orientation is necessary. The two papers in this section should give a good overview on what we know and what we suspect on chromosomes and behavior.

Typically, chromosomal studies have related types and numbers of chromosomes to behavioral patterns. In such a "chromosome-behavior axis" little is known about (a) the specific gene or concise portion of chromosomes that are related to behavioral patterns, (b) the intermediary biochemical steps between chromosome constitution and behavior, and (c) influence of the environment on this chromosome-behavior axis. In a sense we are not much further along than the "taxonomic" phase of these studies. Hopefully, we shall soon enter a much deeper understanding of the relationship.

With these limitations in mind, let us examine initially a research paper relating chromosomes to behavioral consequences. Telfer and her associates examined all male criminals in a number of institutions and found that a disproportionate number, 1 in 11, displayed a gross chromosomal error. The errors were of two kinds—Klinefelter males with an XXY composition and those with an XYY composition (no designated name). The researchers conclude that "gross chromosomal errors contribute small but consistent numbers to the pool of antisocial and aggressive males." As reported by McKusick, 1969, "W. Court Brown suggests that finding the correlation between the XYY karyotype and disturbances of behavior may be the most important discovery yet made in human cytogenetics, (because) it may provide a powerful lever to open up the study of human behavioral genetics."

The Heller review summarizes "where we are today" on this subject. Only a little commentary need be added on the social and legal implications of the YY syndrome to the paper. The Speck case is unclear because of the conflicting reports of his chromosomal characteristics. Even so, the legal profession must now deal with the difficult legal issues of the responsibility of an individual having an XXY or XYY chromosomal syndrome. Professor John Flackett of the Boston College Law School has assembled a small bibliography on the legal implications of the XYY chromosome (provided by Professor Charles H. Baron of that institution). The following titles (no references) are not exhaustive but they do provide an indication of how law journals are responding to the issues engendered by chromosomal abnormalities.

a) The XYY syndrome: genetics, behavior, and the law.
b) The XYY syndrome: its effect on criminal responsibility in New York.
c) Will the XYY syndrome abolish guilt?
d) The XYY chromosome defense.
e) The XYY syndrome: a challenge to our system of criminal responsibility.

Additional Reading

Report on the XYY chromosomal abnormality issued by the National Institute of Mental Health, October 1970, PHS Publication No. 2103, U.S. Government Printing Office. An excellent group of researchers reached the following conclusions (a) the prevelance rates of the XYY chromosome anomaly was much higher for a person in penal and mental institutions than for the general population, (b) definite causal links between XYY chromosome complement and deviant criminal or violent behavior cannot be established, (c) many XYY individuals display no behavioral abnormalities, (d) behavioral aberrations thus far do not indicate a direct cause and effect relationship with the XYY chromosome constitution, and (e) social judgments about XYY individuals based upon his chromosomal constitution are at present unjustified.

Questions

The Heller article has a number of questions on pages 73-74 which the reader might consider.

Also consider the following:

Assuming that (a) the Heller paper is an accurate review of the current knowledge of chromosomes and behavior today and (b) that you are a state legislator, what state bills would you file, if any, on the state's criminal code?

6. Incidence of Gross Chromosomal Errors Among Tall, Criminal American Males

MARY A. TELFER, GERALD R. CLARK, DAVID BAKER, AND CLAUDE E. RICHARDSON

The prevalence of aneuploidy* among criminal males who are mentally ill, mentally retarded, or criminally insane is a phenomenon well appreciated in Great Britain [1, 2] but little recognized in the United States. In the course of a recent search for 47,XYY males among several criminal populations in Pennsylvania, we were impressed by the fact that 1 in 11 tall males displayed a gross chromosomal error, but that all the affected individuals had gone undiagnosed despite frequent arrest and review (Table 1).

Table 1. Incidence of gross chromosomal errors among tall, criminal American males, 71 in. or more in height*

		No. of subjects with chromosomal disorders		
Type of facility	N	Klinefelter males	47,XYY	Overall incidence
JD	14	0	1	1:14
MDDA	30	2	0	1:15
UDA	35	1	2	1:12
CI	50	4	2	1:8
Total				1:11†

*This table appeared on page 1249 in the original article.
†Probability that this incidence is due to chance alone, $p = 0.001$.
Abbreviations: N, number of subjects studied; JD, detention center for juvenile delinquents; MDDA, penal institution for mentally defective delinquent adults; UDA, penal institution for unselected delinquent adults; and CI, mental hospital for the criminally insane.

*An unbalanced set of chromosomes [Ed. note].

Reprinted from *Science,* 1968, 159, 1249-1250, with permission of the Journal and Dr. Telfer.

As a first step, inmates of four institutions for the detention of criminals were screened according to height, those 71 in. tall or over being selected for study. With the explicit permission of the subject, a buccal smear was made according to the method of Sanderson and Stewart [3] and 2 ml of venous blood was drawn into a heparinized syringe for the purpose of leukocyte culture [4]. Two culture vials were set up for each subject. Chromosome counts were made of 25 well-spread metaphases; the unique morphology of the human Y chromosome makes the identification of XYY males by microscopic inspection quite satisfactory. Six clear metaphases were photographed on 4- by 5-in. (10- by 12-cm) film and karyotypes were constructed for each aneuploid subject. Individuals whose cultures failed were eliminated from the study. Mosaicism was not observed but cannot be ruled out as a possibility without parallel analyses of other tissues.

Individuals who displayed sex chromatin in the buccal smear or who demonstrated aneuploidy on chromosomal analysis were revisited for further study. The cytogenetic studies were repeated; in each instance the initial finding was confirmed. A physical examination was performed at this time and the prisoner's social, educational, and medical records were carefully reviewed.

As shown in Table 1, 1 in 11 subjects proved to be aneuploid. Seven of the 129 subjects were Klinefelter males with positive sex chromatin and palpably atrophic tests. Five others were 47,XYY males, including one Negro, apparently the first to be reported in the literature [5].

The incidence of gonosomal aneuploidy among tall American males in a facility for the detention of juvenile delinquents proved to be 1:14; in a penal institution for mentally defective delinquent adults, 1:15; in a penal institution for unselected delinquent adults, 1:12; and in a mental hospital for the criminally insane, 1:8. The comparable incidence of sex chromosome errors among tall men at large is estimated to be 1:80, if one assumes that 20% of American males attain a height of 6 ft or over [5], that sex chromosome errors result in extreme body height, and that the incidence of 47,XYY is 1:2000 [6] and of 47,XXY is 1:500 [7] adult males.

The results of this limited survey appear to confirm British observations that gross chromosomal errors contribute, in small but consistent numbers, to the pool of antisocial, aggressive males who are mentally ill and who become institutionalized for criminal behavior. Our data show, furthermore, that these men are to be found in general prisons as well as in mental hospitals for the "hard to handle."

To this we would add the observation that, despite good physical care and much psychiatric attention throughout repeated incarcerations, these individuals are not being identified in the institutions we have surveyed. The implications of gross chromosomal errors for the intellectual, emotional, physical, and social development of the individual, for his legal status before the law [8], for the psychiatrist who treats him, for the society that must provide either care or parole, are fundamental and deserve serious attention by professionals in many related disciplines.

Bibliography

1. Forssman, H., and G. Hambert. *Lancet,* **1963–I,** 1327.

2. Jacobs, P. A., M. Brunton, M. M. Melville, R. P. Brittain, and W. F. McClemont. *Nature,* 1965, **208,** 1351.

3. Sanderson, A. R., and J. S. S. Stewart. *Brit. Med. J.,* 1961, **2,** 1065.

4. Arakaki, D. T., and R. S. Sparkes. *Cytogenetics,* 1963, **2,** 57.

5. Welch, J. P., D. S. Borgaonkar, and H. M. Herr. *Nature,* 1967, **214,** 500.

6. Anonymous. *Lancet,* **1966–I,** 583.

7. Maclean, N., D. G. Harnden, W. M. Court Brown, J. Bond and D. J. Mantle. *Lancet,* **1964–I,** 286.

8. Pritchard, M. *Lancet,* **1964–II,** 762.

7. Human Chromosome Abnormalities as Related to Physical and Mental Dysfunction

JOHN H. HELLER

The relationship of human disease syndromes to chromosome aberrations is assuming an increasingly greater role in the detection, diagnosis, treatment and prediction of mental and physical defects in man. By means of karyotype analysis one is enabled to recognize previously unknown syndromes and to differentiate between separate but phenotypically similar entities. Proper diagnosis permits suitable therapeutic measures to be undertaken and enables genetic counselors to assess correct risks in many instances. Recent refinements in sampling embryonic cells by amniocentesis make it feasible to determine, in high risk cases, whether the embryo has a chromosome abnormality or whether it is a male, which has a high risk of sex-linked genetic defect. Termination of pregnancy can be recommended on the basis of this knowledge.

Classes of Chromosome Abnormalities

Chromosome abnormalities have been known in plant and animal species for a very long time. They occur firstly as variations in the number of chromosomes per cell deviating from the normal two sets (maternal and paternal), existing either as complete multiples of sets, a condition called polyploidy (triploidy, tetraploidy, etc.), or as addition or loss of chromosomes within a set, a situation known as aneuploidy (monosomy, trisomy, tetrasomy, etc.). The origin of deviations in chromosome number is known to be through nondisjunction, either during the meiotic divisions in the maturation of the germ cells or during mitotic divisions in the developing individual, or through lagging of chromosomes at anaphase of cell division.

Secondly, chromosome aberrations occur as structural modifications such as duplications, deficiencies, translocations, inversions, isochromosomes, ring chromosomes, etc. These aberrations result from chromosome breakage and reunion in various patterns different from the normal sequence of loci. In most cases, especially the "spontaneous" instances, the cause of chromosome breaks is unknown, but many extra-

Reprinted by permission from *Journal of Heredity,* 1969, Volume 60, 239-248.

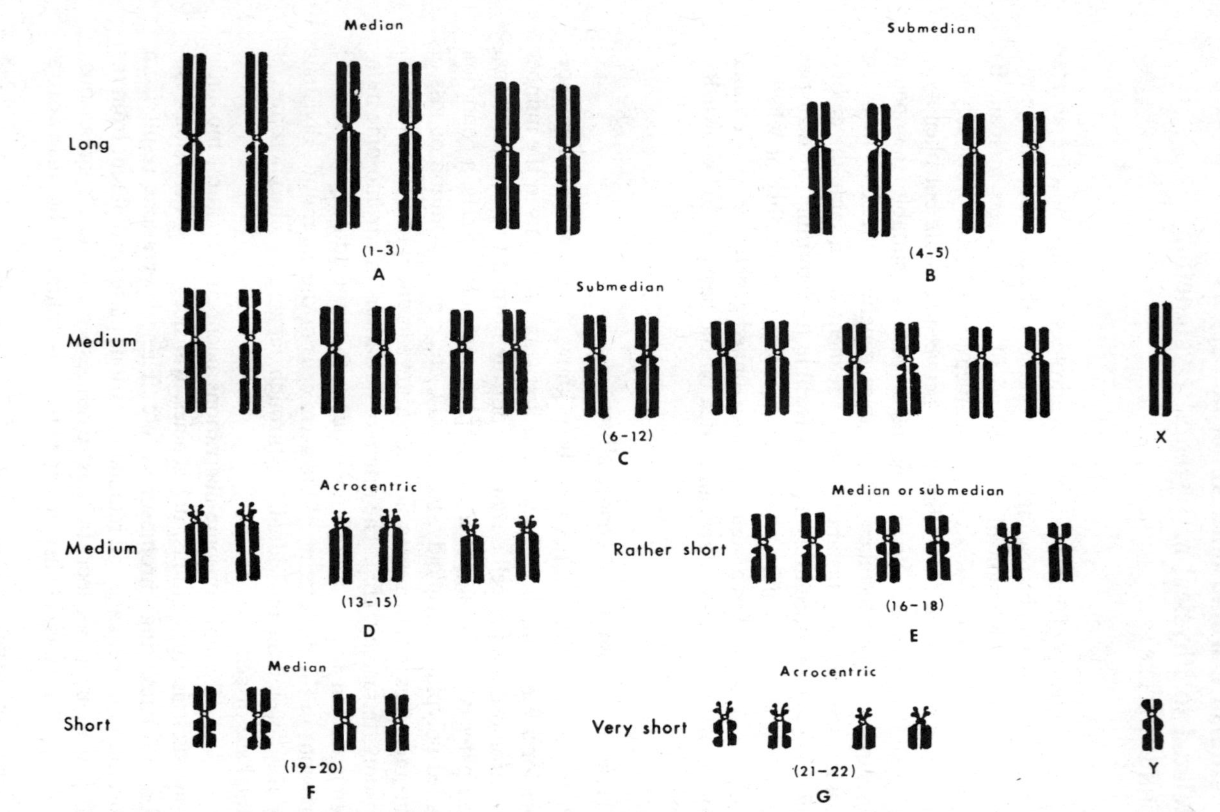

Fig. 1 Idiogram of normal male with 22 pairs of autosomes and XY sex chromosome constitution (modified from Patau[69], Sohval[84], Ferguson-Smith *et al.*[27], and Palmer and Funderburk[68a]).

neous agents have been demonstrated experimentally to be efficacious in inducing fragmentation. Foremost among these agents is ionizing radiation but many chemical substances (alkylating agents, nitroso-compounds, antibiotics, DNA precursors, etc.) and viruses have been implicated.

Genetic Effects of Chromosome Aberrations

The striking genetic alterations accompanying chromosome aberrations were brilliantly analyzed by Blakeslee and coworkers on *Datura,* and by the *Drosophila* workers (Morgan, Bridges, Múller, Sturtevant, Painter, Patterson and many others). The task was greatly facilitated in *Drosophila* by the fortunate circumstance in the larval salivary glands where the giant polytene chromosomes exhibit intimate somatic pairing as well as characteristic banding patterns that permit identification of specific gene loci.

Particularly illuminating were Bridges' analyses of sex chromosomes and sex determination in *Drosophila,* utilizing the phenomenon of nondisjunction of the sex chromosomes and culminating in the genic balance theory of sex determination. In this insect the female normally has two X chromosomes plus the autosomes, and the male has one X and one Y. Two X chromosomes and one Y chromosome results in a female, whereas a chromosome constitution of XO produces a sterile male.

In contrast, the Y chromosome in mammals has a strongly masculinizing influence. The presence of a single Y is sufficient to induce differentiation into a male phenotype in the presence of one to five X chromosomes. The XO constitution differentiates into a female phenotype in both mouse and man.

Mammalian Chromosome Studies

The first reported instance of chromosome aberration in mammals was discovered by genetic methods in the waltzing mouse by William H. Gates in 1927[37], and analyzed cytologically by T. S. Painter[68]. Many difficulties in techniques prevented accurate counting and analysis of mammalian chromosomes—large number and relative small size of chromosomes, tendency to clump on fixation, cutting of chromosomes in sectioned material, etc. Even the somatic chromosome number in man was accepted erroneously as 48 until 1956 when Tjio and Levan [88] established the correct count of 46. This count was quickly confirmed by

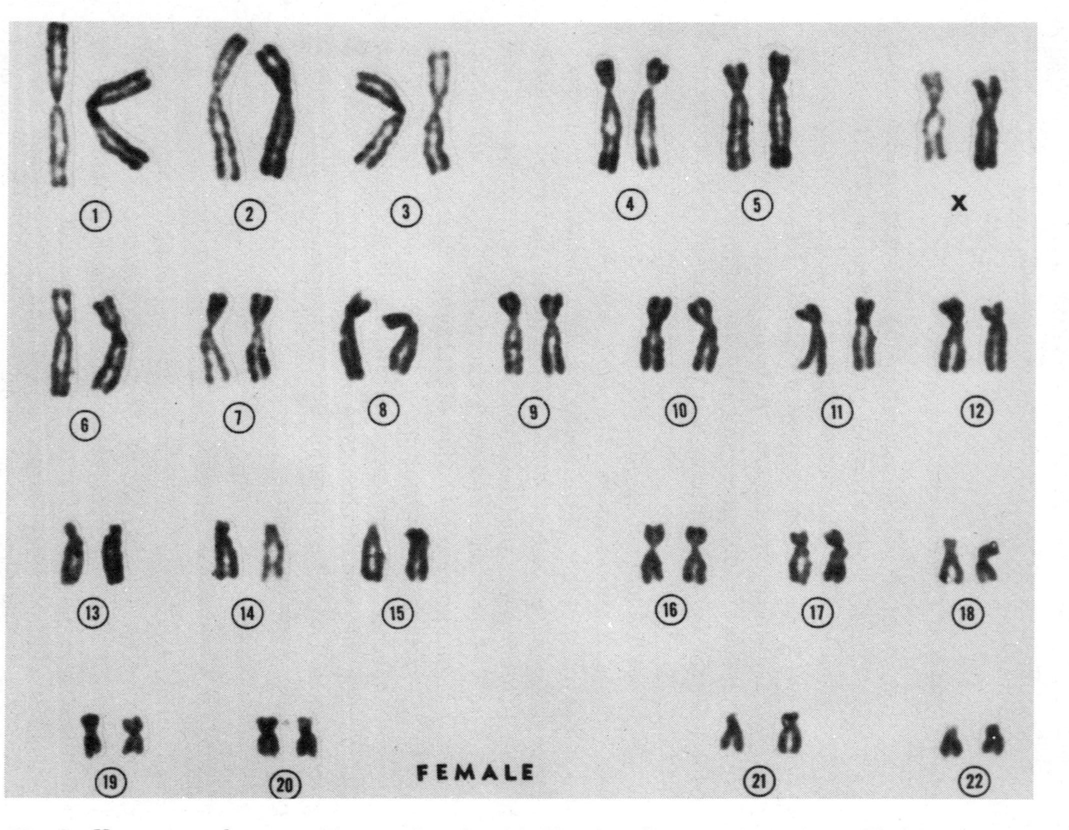

Fig. 2 Karyotype of a normal human female with 22 pairs of autosomes and two X chromosomes.

Ford and Hammerton[28], and in 1959 the first positive correlation of a chromosome abnormality and human disease syndrome was made by Lejeune *et al.*[54] (also Jacobs *et al.*[44])—the trisomic number 21 chromosome, and Down's syndrome or mongolism. Shortly thereafter Klinefelter's[46] and Turner's[30] syndromes were identified with XXY and XO sex chromosome constitutions respectively, and in rapid succession reports of many other human chromosome abnormalities appeared, such as trisomy 17, trisomy 18, partial trisomy, ring X chromosome, sex chromosome mosaics, cri-du-chat syndrome, et cetera[9, 26].

This sudden explosion of human chromosome studies, in contrast to the long delay of confirmation in human cells of chromosome abnormalities long known in plants and other animals, was made possible by new techniques of preparation. The accumulation of many cells in the metaphase stage of mitosis with colchicine, the use of hypotonic solution to swell the cells and separate chromosomes on the spindle, the discovery that phytohemagglutinin stimulates mammalian peripheral lymphocytes to undergo mitosis, and the method of squashing or spreading on slides of loose cells taken from bone marrow or tissue culture, all contributed to the rapid and accurate analysis of mammalian and human chromosome number and structure.

Karyotype analysis involves the careful comparison of chromosomes in a particular individual to the standard pattern for human cells, including precise measurements of lengths, arm ratios and other morphological features. Special attention is given to comparison of homologous chromosomes where differences may indicate abnormalities. An idiogram is a diagrammatic representation of the entire standard chromosome complement, showing their relative lengths, position of centromeres, arm ratios, satellites, secondary constrictions and other features. Figure 1 shows an idiogram of a normal human male with 22 pairs of autosomes and XY sex chromosome constitution. A karyotype is constructed from photographs of chromosomes which are arranged in pairs similar to the idiogram. Figure 2 shows a karyotype of a normal human female.

Incidence of Human Chromosome Anomalies

Chromosome anomalies are relatively frequent events. They have been estimated to occur in 0.48 percent of all newborn infants (one in 208)[81]. At least 25 percent of all spontaneous miscarriages result from gross chromosomal errors[13]. The general incidence of chromosome abnormalities in abortuses is more than fifty times the incidence at birth.

Although it is impossible to obtain an accurate total of victims suffering from effects of chromosome aberrations, one can make rough calculations on the basis of their estimated frequencies in the population of the United States assuming that there is no appreciable difference in life expectancy between these individuals and those with normal chromosome complements. Although this assumption probably is unjustifiable, it suffices for this rough calculation. Among the current population of 202 million we arrive at a figure of 1,136,971 total afflicted with chromosome abnormalities. This total probably represents an underestimate since it does not include all types of chromosome aberrations. Table 1 indicates totals for a number of specific syndromes.

Table 1. Total frequencies in the United States of various types of chromosomal abnormalities, calculated on the basis of 202 million current population and the estimated frequency of each abnormality. (It must be noted that the grand total does not include all types of chromosome aberrations, therefore must be lower than the real value)

Syndrome	Chromosome number	Estimated incidence	Calculated number in U.S.
Down's trisomy 21	47	1 in 700	288,571
Trisomy D	47	1 in 10,000	20,200
Trisomy E	47	1 in 4000	50,500
Trisomy X	47	1 in 10,000 females	101,000
Turner's XO	45	1 in 5000 females	20,200
Klinefelter's XXY	47	1 in 400 males	252,500
Double Y XYY	47	1 in 250 males	404,000
		Total	1,136,971

Syndromes Related to Autosome Abnormalities

Down's syndrome

This defect results from duplication of all or part of autosome 21, either in the trisomic state or as a translocation to another chromosome, usually a 13-15 (D group) or 16-18 (E group) but may be to another G group chromosome. The overall incidence is about 1 in 700 live births[71], but the trisomic type is correlated with age of the mother, having a frequency of about 1 in 2000 in mothers under 30 years of age, and increasing to 1 in 40 in mothers aged 45 or over. The translocation type constitutes about 3.6 percent of cases and is unrelated to the mother's age, but is transmitted in a predictable manner. Among mental retardates mongoloids represent 16.7 percent.

Clinical features include physical peculiarities ranging from slight anomalies to severe malformations in almost every tissue of the body. Typical appearance of a mongoloid shows slanting eyes, saddle nose, often a large ridged tongue that rolls over a protruding lip, a broad, short skull and thick, short hands, feet and trunk. Frequent complications occur: cataract or crossed eyes, congenital heart trouble, hernias, and a marked susceptibility to respiratory infections. They exhibit characteristic dermatoglyphic patterns on palms and soles. Also they have many biochemical deviations from normal, such as decreased blood-calcium levels and diminished excretion of tryptophane metabolites. Early ageing is common.

All mongoloids are mentally retarded; they usually are 3 to 7 years old mentally. Among the relatively intelligent patients, abstract reasoning is exceptionally retarded.

Female mongoloids are fertile and recorded pregnancies have yielded approximately 50 percent mongoloid offspring. Fortunately male mongoloids are sterile. Examination of their testes reveals varying degrees of spermatogenic arrest correlated with the abnormal chromosome features.

Among mongoloids there is a prevelence of leukemia in childhood; the incidence is some twenty times greater than in the general population. Simultaneous occurrence with other syndromes such as Klinefelter's, also is found, and many cases of mosaicism have been described.

E trisomy syndrome

This is another autosomal anomaly, which involves chromosomes 16, 17 and 18, and is estimated to occur at a frequency of 1 in 4000 live births[20]. Many others die before birth, thus contributing to the large number of miscarriages and stillbirths. These individuals survive only a short time, from one-half day to 1460 days, with an average of 239 days, but females live significantly longer than males.

Trisomy 17 syndrome

Many serious defects usually are present in afflicted individuals[25]: odd shaped skulls, low-set and malformed ears, triangular mouth with receding chin, webbing of neck, shield-like chest, short stubby fingers, and toes with short nails, webbing of toes, ventricular septal defect and mental retardation, as well as abnormal facies, micrognathia and high arched palate.

Trisomy 18 syndrome

This anomoly[70, 82] is characterized by multiple congenital defects of which the most prominent clinical features are: mental retardation with moderate hypertonicity, low-set malformed ears, small mandible, flexion of fingers with the index finger overlying the third, and severe failure to thrive. It generally results in death in early infancy. Its frequency increases with advanced maternal age. Three times as many females as males have been observed; one would expect that more males with this syndrome will be found among stillbirths and fetal deaths.

D syndrome

This trisomy [19, 56, 70, 83] involves chromosomes 13, 14 and 15, and has an estimated frequency of about 1 in 10,000 live births. Many others die in utero. Survival time has been reported from 0 to 1000 days, with an average of 131 days.

Clinical features include: microcephaly, eye anomalies (corneal opacities, colobomata, microphthalmia, anophthalmia), cleft lip, cleft palate, brain anomalies (particularly arrhinencephaly), supernumerary digits, renal anomalies (especially cortical microcysts), and heart anomalies.

Trisomy 22 syndrome

This syndrome produces mentally retarded, schizoid individuals. Reports of its occurrence are too few to permit an estimate of its frequency in the population.

Cri-du-chat syndrome

Lejeune *et al.*[55] first described this anomaly in 1965, which involves a deficiency of the short arm of a B group chromosome, number 5. Translocations appear to be a common cause of the defect, an estimated 13 percent of cases being associated with translocations; described cases have had B/C, B/G, and B/D translocations[23]. The high proportion emphasizes the importance of unbalanced gamete formation in translocation heterozygotes as a cause of this syndrome. Among parents the frequency of male and female carriers is approximately the same, a situation that contrasts with the much greater frequency of female carriers of a D/G translocation among parents of translocation mongoloids.

Typical clinical features of cri-du-chat individuals are: low birth weight, severe mental retardation, microcephaly, hypertelorism, retrognathism, downward slanting eyes, epicanthal folds, divergent strabismus, growth retardation, narrow ear canals, pes planus and short metacarpals and metatarsals. About 25 to 30 percent of them have congenital heart disorders. A characteristic cat-like cry in infancy is responsible for the name of the syndrome. The cry is due to a small epiglottis and larynx and an atrophic vestibule. However, this major diagnostic sign disappears after infancy, making identification of older cases difficult.

An estimate of the frequency of this syndrome is given as over 1 percent but less than 10 percent of the severely mentally retarded patients. Many have IQ scores below 10, and most are institutionalized.

Philadelphia chromosome

Finally, among autosomal aberrations, a deleted chromosome 21 occurs in blood-forming stem cells in red bone marrow. This deletion, which shows up long after birth, appears to be the primary event causing chronic granulocytic leukemia. This aberration was discussed in 1960 by Nowell and Hungerford[67] (also Baikie *et al.*[3]).

Syndromes Related to Sex Chromosome Aberrations

The great majority of known chromosomal abnormalities in man involve the sex chromosomes. In one survey (that excluded XYY) it was estimated that abnormalities occurred in 1 out of every 450 births; if the recent estimate of XYY[81] is correct, the frequency actually is much higher. Increased knowledge about sex chromosome aberrations is probably related to the greater concentration of attention on patients with sexual disorders, but is due in part to the ability to detect carriers of an extra X chromosome by the so-called sex chromatin body or Barr body[6]. This structure is a stainable granule at the periphery of a resting nucleus and, according to the Lyon hypothesis[57], is considered to be an inactivated X chromosome. A normal femal cell has one Barr body, since it has two X chromosomes, and is said to be sex chromatin positive (or one positive). A normal male cell has no Barr body and is said to be sex chromatin negative.

Klinefelter's syndrome

The first sex chromosome anomaly described in 1959 by Jacobs and Strong[46] and also by Ford *et al.*[29] was the XXY constitution that is typical of Klinefelter's syndrome. Buccal smears from these patients are sex chromatin positive. They can be tentatively diagnosed by this test along with clinical symptoms. Final confirmation of diagnosis can be achieved by karyotype analysis using either bone marrow aspiration or peripheral blood culture.

Victims of Klinefelter's syndrome are always male but they are generally underdeveloped, eunachoid in build, with small external genitalia, very small testes and prostate glands, with underdevelopment of hair on the body, pubic hair and facial hair, frequently with enlarged breasts (gynecomastia), and many have a low IQ.

The classical type with two X chromosomes and one Y chromosome was the first case discovered, but subsequently chromosome compositions of XXXY, XXXXY, XXYY[66] and XXXYY[7, 8, 63, 77] have been reported. In addition, numerous mosaics have been described, including double, triple and quadruple numeric mosaics, as well as combinations of numeric and structural mosaics. These conditions are summarized in Table 2. They all resemble the XXY Klinefelter's phenotypically and are considered modified Klinefelter's syndromes. The classical XXY type may have low normal mental development or may be retarded, but other types show increasingly greater mental retardation.

The incidence of Klinefelter's syndrome is estimated to be 1 in 400 male live births, which represents from 1 to 3 percent of mentally deficient patients. This condition also has been correlated with age of the mother: the older the mother, the greater the risk of having such a child. These individuals usually are sterile. Spermatogenesis is generally totally absent. Hyalinization of the semeniferous tubules begins shortly before puberty. Congenital malformations are rare. Mental retardation is present in approximately 25 percent of affected individuals, and mental illness may be more common than in the general population.

Turner's syndrome

Female gonadal dysgenesis was described by Turner in 1938 as a syndrome of primary amenorrhea, webbing of the neck, cubitas valgus and

Table 2. Reported sex chromosomal constitutions in Klinefelter's syndrome (modified from Reitalu[77])

		Sex chromosomal constitution			
Only one karyotype observed per individual		XXY			
		XXYY			
		XXXY			
		XXXYY			
		XXXXY			
Numeric mosaics	Double	XX	XXY		
		XY	XXY		
		XY	XXXY		
		XXY	XXYY		
		XXXY	XXXXY		
		XXXX	XXXXY		
	Triple	XY	XXY	XXYY	
		XX	XXY	XXXY	
		XY	XXY	XXXY	
		XO	XY	XXY	
		XX	XY	XXY	
		XXXY	XXXXY	XXXXYY	
		XXXY	XXXXY	XXXXXY	
	Quadruple	XXY	XY	XX	XO
Numeric and structural mosaics	Double	XXY	XXxY		
	Triple	XY	XXY	XXxY	
		XxY	Xx	XY	

short stature, coarctation of aorta, failure of ovarian development and hormonal abnormalities. Patients exhibit sexual infantilism; their breasts are usually underdeveloped, nipples often widely spaced, particularly in those subjects who have a shield or funnel chest deformity. Usually sexual hair is scanty; external genitalia are infantile; labia small or unapparent; clitoris usually normal, although may be enlarged. The uterus is infantile; the tubes long and narrow; the gonads represented by long, narrow, white streaks of connective tissue in normal position of ovary. They are almost always sterile. Hormonal secretions usually are abnormal. Shortness of stature is characteristic and many other skeletal abnormalities occur. Peculiar facies include small mandible, anti-mongolian slant of eyes, depressed corners of mouth, low-set ears, auricles sometimes deformed. Cardiovascular defects are frequent, the most common being coarctation of the aorta. Slight intellectual impairment is found in some patients, particularly those with webbing of the neck.

In 1954 it was discovered that many patients with ovarian agenesis were sex chromatin negative, and in 1959 Ford and colleagues[30] gave the first chromosome analysis showing that Turner's syndrome has the sex chromosome abnormality of only one X chromosome (XO) rather than two X's. It was quickly confirmed by Jacobs and Keay[45] and by Fraccaro *et al*[33].

Mosaicism is known to exist—both 45 chromosome cells and 46 chromosome cells occur side by side in tissues of the individual—and can result from nondisjunction in early embryonic development. Isochromosomes sometimes are involved, e.g., creating a situation with 3 long arms of the X chromosome but only 1 short arm.

The incidence of XO Turner's syndrome is estimated as 1 in approximately 5000 women; many die in utero.

Table 3. The relationship between sex, sex chromosome complement and sex chromatin pattern (modified from Miller[63])

Sex chromatin pattern	Sexual phenotype	
	Female	Male
–	XO	XY
–	XY (testicular feminization)	XYY
+	XX	XXY
		XXYY
++	XXX	XXXY
		XXXYY
+++	XXXX	XXXXY
++++	XXXXX	

Large scale screening of newborn babies by buccal smears can permit detection of chromatin negative females, chromatin positive males, and double, triple, quadruple and quintuple positive cases of either sex[58]. Table 3 shows the relationship between sex chromosome complements and sex chromatin pattern.

Triplo-X syndrome

Females containing three[47], four and five X chromosomes are known[4, 47, 63]. The triplo-X syndrome is thought to have an incidence of about 1 in 800 live female births. This syndrome was first described by Jacobs *et al.*[47] in 1959. Although it has no distinctive clinical picture, menstrual irregularities may be present, secondary amenorrhea or premature menopause. Most cases have no sexual abnormalities and many are known to have children. The most characteristic feature of 3X females is mental retardation. Quadruple-[14] and quintuple-X[50] syndromes are much rarer. These individuals are mentally retarded, usually the more X chromosomes present, the more severe the retardation. Frequently these individuals are fertile.

An extra X chromosome confers twice the usual risk of being admitted to a hospital with some form of mental illness. The loss of an X, on the other hand, has no association with mental illness; thus the chance of mental hospital admission is not raised for an XO female. An extra X chromosome also predisposes to mental subnormality. The prevalence of psychosis among patients in hospitals for the subnormal is unusually high in males with two or more X chromosomes.

Numerous other sex chromosome anomalies occur[38], many involving mosaics and structural chromosome aberrations. For example, occasionally an XY embryo will differentiate into a female, a situation referred to as testicular feminization male pseudohermaphrodite (Morris syndrome)[60]. These individuals have only streak gonads and vestigial internal genital organs. They usually have undeveloped breasts and do not menstruate. They are invariably sterile[76].

Still other sexual abnormalities are intersexes and true hermaphrodites, many of which have an XX sex chromosome constitution or are mosaics for sex chromosomes such as XO/XY or XX/XY or more complicated mixtures[31]. Sex chromosome mosaicism is very common. Almost every sex chromosome combination found alone has been found in association with one or more cell lines with a different sex chromosome constitution. These mosaics exhibit quite a variable expression; for example in an XO/XY mosaic the external genitalia can appear female, male or intersexual[85].

The YY syndrome

The male with an extra Y chromosome (XYY) has attracted much attention in the public press as well as in scientific circles because of his reputed antisocial, aggressive and criminal tendencies[1, 2, 64]. Although this abnormality belongs in the above category of syndromes related to sex chromosome aberrations, it has been singled out for special discussion because of its social and legal implications.

Evidence supporting the existence of a double Y syndrome has accumulated within the last six years. Studies in Sweden[32] showed an unusually large number of XXYY and XYY men among hard-to-manage patients in mental hospitals. These observations received impressive confirmation in studies of maximum security prisons and hospitals for the criminally insane in Scotland where an astonishingly high frequency (2.9 percent) of XYY males were found[48]. This was over fifty times higher than the then current estimate of 1 in 2000 in the general population. Subsequently many additional studies on the YY syndrome have appeared and a composite picture of the XYY male emerged[5, 19, 15, 21a, 34, 35, 41, 72-75, 78, 80, 92].

The principal features of the extra Y syndrome appear to be exceptional height and a serious personality disorder leading to behavioral disturbances. It seems likely it is the behavior disorder rather than their intellectual incompetence that prevents them from functioning adequately in society[18].

Clinically the XYY males are invariably tall (usually six feet or over) and frequently of below-average intelligence. They are likely to have unusual sexual tastes, often including homosexuality. A history of antisocial behavior, violence and conflict with the police and educational authorities from early years is characteristic[86] of the syndrome.

Although these males usually do not exhibit obvious physical abnormalities[12, 24, 40, 42, 52, 91], several cases of hypogonadism[11], some with undescended testes, have been reported. Others have epilepsy, malocclusion and arrested development[87], but these symptoms may be fortuitously associated. One case was associated with trisomy 21[61], another with pseudohermaphrodism[35]. The common feature of an acne-scarred face may be related to altered hormone production. The criminally aggressive group were found to have evidence of an increased androgenic steroid production as reflected by high plasma and urinary testosterone levels[12, 43]. If the high level of plasma testosterone is characteristic of XYY individuals, it suggests a mechanism through which this condition may produce behavioral changes, possibly arising at puberty.

Antisocial and aggressive behavior in XYY individuals may appear early in life, however, as evidenced by a case reported by Cowie and

Kahn[22]. A prepubertal boy with normal intelligence, at the age of 4½ years, was unmanageable, destructive, mischievous and defiant, overadventurous and without fear. His moods alternated; there were sudden periods of overactivity at irregular intervals when he would pursue his particular antisocial activity with grim intent. Between episodes he appeared happy and constructive. The boy was over the 97th percentile in height for his age, a fact that supports the view that increased height in the XYY syndrome is apparent before puberty.

It has been suggested that the ordinary degree of aggressiveness of a normal XY male is derived from his Y chromosome, and that by adding another Y a double dose of those potencies may facilitate the development of aggressive behavior[65] under certain conditions. A triple dose (XYYY) would be present in the case reported by Townes *et al.*[90].

The first reported case of an XYY constitution[39, 79] was studied because the patient had several abnormal children, although he appeared to be normal himself. Until recently, reports of the XYY constitution have been uncommon, probably because no simple method exists for screening the double-Y condition that is comparable to the buccal smear—sex chromatin body technique for detecting an extra X chromosome. Another possible explanation for the rarity of reports on the XYY karyotype is the absence of a specific phenotype in connection with it. Most syndromes with a chromosome abnormality are ascertained because of some symptom or clinical sign that indicates a need for chromosome analysis. Consequently there have been few studies that place the incidence of this chromosome abnormality in its proper perspective to the population as a whole.

Very recently a study of the karyotypes of 2159 infants born in one year was made by Sergovich *et al.*[81]. These investigators detected 0.48 percent of gross chromosome abnormalities. In this sample the XYY condition appeared in the order of 1 in 250 males, which would make it the most common form of aneuploidy known for man. The previous estimate was about 1 in 2000 males. If this figure of 1/250 is valid for the population as a whole, it means that the great majority of cases go undetected and consequently must be phenotypically normal and behave near enough to the norm to go unrecognized.

Several cases of asymptomatic males have been published, including the first one described (Sandberg *et al.*[79] and Hauschka *et al.*[39]), which proved to be fertile. It appears that the sons of XYY men do not inherit their father's extra Y chromosome[59a].

Another fertile XYY male, reported by Leff and Scott[53], had inferiority feelings, was slightly hypochondriacal and obsessional, and not very aggressive. He gave a general impression of emotional immaturity. He was 6 feet, 6 inches tall, healthy, with normal genitalia and electroencephalogram. His IQ was 118. Wiener and Sutherland[93] discovered by chance an XYY male who was normal; he was 5 feet, 9½ inches tall, with normal genitalia and body hair, normal brain waves, and with an IQ of 97. He exhibited a cheerful disposition and mild temperament, had no apparent behavioral disturbance and never required psychiatric advice. This case supports the idea that an XYY male can lead a normal life.

Social and Legal Implications of the YY Syndrome

The concept that when a human male receives an extra Y chromosome it may have an important and potentially antisocial effect upon his behavior is supported by impressive evidence[15, 21]. Lejuene states that "There are no born criminals but persons with the XYY defect have considerably higher chances." Price and Whatmore[74] describe these males as psychopaths, "unstable and immature, without feeling or remorse, unable to construct adequate personal relationships, showing a tendency to abscond from institutions and committing apparently motiveless crimes, mostly against property." Casey and coworkers[16] examined the chromosome complements in males 6 feet and over in height and found: 12 XYY among 50 mentally subnormal and 4 XYY among mentally ill patients detained because of antisocial behavior; also 2 XYY among 24 criminals of normal intelligence. They concluded that their results indicate that an extra Y chromosome plays a part in antisocial behavior even in the absence of mental subnormality. The idea that criminals are degenerates because of bad heredity has had wide appeal. There is no doubt that genes do influence to some extent the development of behavior. The influence may be strongly manifested in some cases but not in others. Some individuals appear to be driven to aggressive behavior.

Several spectacular crime cases served to publicize this genetic syndrome, and it has been played up in newspapers, news magazines, radio and television. In 1965 Daniel Hugon, a stablehand, was charged with the murder of a prostitute in a cheap Paris hotel. Following his attempted suicide he was found to have an XYY sex chromosome constitution. Hugon surrendered to the police and his lawyers contended that he was unfit to stand trial because of his genetic abnormality. The prosecution asked for five to ten years; the jury decided to give him seven.

Richard Speck, the convicted murderer of eight nurses in Chicago in 1966, was found to have an XYY sex chromosome constitution. He has all the characteristics of this syndrome found in the Scottish survey: he is 6 feet 2 inches tall, mentally dull, being semiliterate with an IQ of 85, the equivalent of a 13-year-old boy. Speck's face is deeply pitted with acne scars. He has a history of violent acts against women. His aggressive behavior is attested by his record of over 40 arrests. Speck was sentenced to death but the execution has been held up pending an appeal of the conviction.

In Melbourne, Australia, Lawrence Edward Hannell, a 21-year-old laborer on trial for the stabbing of a 77-year-old widow, faced a maximum sentence of death. He was found to have an XYY constitution, mental retardation, an aberrant brain wave pattern, and a neurological disorder. Hannell pleaded not guilty by reason of insanity, and a criminal court jury found him not guilty on the ground that he was insane at the time of the crime.

A second Melbourne criminal with an XYY constitution, Robert Peter Tait, bludgeoned to death an 81-year-old woman in a vicarage where he had gone seeking a handout. He was convicted of murder and sentenced to hang, but his sentence was commuted to life imprisonment.

Another case is that of Raymond Tanner, a convicted sex offender, who pleaded guilty to the beating and rape of a woman in California. He is 6 feet 3 inches tall, mentally disordered, and has an XYY complement. A superior court judge is attempting to decide whether Tanner's plea of guilty to assault with intent to commit rape will stand, or whether he will be allowed to plead innocent by reason of insanity.

Criminal lawyers in the United States have already begun to request genetic studies of their clients. In October of 1968 a lawyer for Sean Farley, a 26-year-old XYY man in New York who was charged with a rape-slaying, maneuvered to raise the issue of his client's genetic defect in court.

Many questions are raised by the double Y syndrome—basic social, legal and ethical questions—which will become more and more insistent as the implications of chromosome abnormalities take root in the public mind. Is an extra Y chromosome causally related to antisocial behavior? Is there a genetic basis for criminal behavior? If a man has an inborn tendency toward criminal behavior, can we fairly hold him legally accountable for his acts? If a criminal's chromosomes are at fault, how can we rehabilitate him?

The evidence to date is inadequate to prove conclusively the validity of the syndrome and convict all of the world's estimated five million XYY males of innate aggressive or criminal tendencies. But if the concept is proved, what then? The first step would seem to be to identify the XYY infants in the general population. This suggests the need for a nationwide program of automatic chromosome analysis of all newborns.

How should society deal with XYY individuals? If they are genetically abnormal, they should not be treated as normal. If the XYY condition dooms a man to a life of crime, he should be restrained but not punished. Mongolism also is a chromosome abnormality, and afflicted individuals are not held responsible for their behavior. Some valuable suggestions on the legal aspects of the double Y syndrome have been published recently by Kennedy McWhirter[59]. Elsewhere, Kessler and Moos[51] claim that definitive concepts relating to the YY syndrome have been accepted prematurely.

If all infants could be karyotyped at birth or soon after, society could be forearmed with information on chromosome abnormalities and perhaps it could institute the proper preventive and other measures at an early age. Although society can not control the chromosomes (at least at the present time) it can do a great deal to change certain environmental conditions that may encourage XYY individuals to commit criminal acts.

The theory that a genetic abnormality may predispose a man to antisocial behavior, including crimes of violence, is deceptively and attractively simple, but will be difficult to prove. Extensive chromosome screening with prospective follow-up of XYY males will be essential to determine the precise behavioral risk of this group. It is by no means universally accepted yet. Many geneticists urge that we should be cautious in accepting the interpretation that the double Y condition is specifically associated with criminal behavior, and particularly so with reference to the medicolegal validity of these concepts.

Literature Cited

1. Anonymous. The YY syndrome. *Lancet* **1**:583-584. 1966.

2. ______. Criminal behavior—XYY criterion doubtful. *Science News* **96**:2. 1969.

3. Baikie, A. G., W. M. Court Brown, K. E. Buckton, D. G. Harnden,

P. A. Jacobs, and I. M. Tough. A possible specific chromosome abnormality in human chronic myeloid leukemia. *Nature* **188:** 1165-1166. 1960.

4. ______, O. Margaret Garson, Sandra M. Weste, and Jean Ferguson. Numerical abnormalities of the X chromosome. *Lancet* **1:** 398-400. 1966.

5. Balodimos, Marios C., Hermann Lisco, Irene Irwin, Wilma Merrill, and Joseph F. Dingman. XYY karyotype in a case of familial hypogonadism. *J. Clin. Endocr.* **26:**443-452. 1966.

6. Barr, M. L. Sex chromatin and phenotype in man. *Science* **130:** 679. 1959.

7. ______ and D. H. Carr. Sex chromatin, sex chromosomes and sex anomalies. *Canad. Med. Assn. J.* **83:**979-986. 1960.

8. ______, D. H. Carr, H. C. Soltan, Ruth G. Wiens, and E. R. Plunkett. The XXYY variant of Klinefelter's syndrome. *Canad. Med. Assn. J.* **90:**575-580. 1964.

9. Belsky, Joseph L. and George H. Mickey. Human cytogenetic studies. *Danbury Hospital Bull.* **1:**19-20. 1965.

10. Boczkowski, K. and M. D. Casey. Pattern of DNA replication of the sex chromosomes in three males, two with XYY and one with XXYY karyotype. *Nature* **213:**928-930. 1967.

11. Buckton, Karin E., Jane A. Bond, and J. A. McBride. An XYY sex chromosome complement in a male with hypogonadism. *Human Chromosome Newsletter* No. 8, p. 11. Dec. 1962.

12. Carakushansky, Gerson, Richard L. Neu, and Lytt I. Gardner. XYY with abnormal genitalia. *Lancet* **2:**1144. 1968.

13. Carr, D. H. Chromosome studies in abortuses and stillborn infants. *Lancet* **2:**603-606. 1963.

14. ______, M. L. Barr, and E. R. Plunkett. An XXXX sex chromosome complex in two mentally defective females. *Canad. Med. Assn. J.* **84:**131-137. 1961.

15. Casey, M. D., C. E. Blank, D. R. K. Street, L. J. Segall, J. H. McDougall, P. J. McGrath, and J. L. Skinner. YY chromosomes and antisocial behavior. *Lancet* **2:**859-860. 1966.

16. ______, L. J. Segall, D. R. K. Street, and C. E. Blank. Sex chromosome abnormalities in two state hospitals for patients requiring special security. *Nature* **209**:641-642. 1966.

17. ______, D. R. K. Street, L. J. Segall, and C. E. Blank. Patients with sex chromatin abnormality in two state hospitals. *Ann. Human Genet.* **32**:53-63. 1968.

18. Close, H. G., A. S. R. Goonetilleke, Patricia A. Jacobs, and W. H. Price. The incidence of sex chromosomal abnormalities in mentally subnormal males. *Cytogenetics* **7**:277-285. 1968.

19. Conan, P. E. and Bayzar Erkman. Frequency and occurrence of chromosomal syndromes. I. D-trisomy. *Am. J. Human Genet.* **18**: 374-386. 1966.

20. ______ and ______. Frequency and occurrence of chromosomal syndromes. II. E-trisomy. *Am. J. Human Genet.* **18**:387-398. 1966.

21. Court Brown, W. M. Sex chromosomes and the law. *Lancet* **2**: 508-509. 1962.

21a. ______. Males With an XYY sex chromosome complement. *J. Med. Genet.* **5**:341-359. 1968.

22. Cowie, John and Jacob Kahn. XYY constitution in prepubertal child. *Brit. Med. J.* **1**:748-749. 1968.

23. deCapoa, A., D. Warburton, W. R. Breg, D. A. Miller, and O. J. Miller. Translocation heterozygosis: a cause of five cases of *cri du chat* syndrome and two cases with a duplication of chromosome number five in three families. *Am. J. Human Genet.* **19**:586-603. 1967.

24. Dent, T., J. H. Edwards, and J. D. A. Delhanty. A partial mongol. *Lancet* **2**:484-487. 1963.

25. Edwards, J. H., D. G. Harnden, A. H. Cameron, V. Mary Crosse, and O. H. Wolff. A new trisomic syndrome. *Lancet* **1**:787-790. 1960.

26. Eggen, Robert R. Chromosome Diagnostics in Clinical Medicine. Charles C. Thomas, Springfield, Ill. 1965

27. Ferguson-Smith, M. A., Marie E. Ferguson-Smith, Patricia M.

Ellis, and Marion Dickson. The sites and relative frequencies of secondary constrictions in human somatic chromosomes. *Cytogenetics* **1**:325-343. 1962.

28. Ford, C. E. and J. L. Hamerton. The chromosomes of man. *Nature* **178**:1020-1023. 1956.
29. ______, K. W. Jones, O. J. Miller, Ursula Mittwoch, L. S. Penrose, M. Ridler, and A. Shapiro. The chromosomes in a patient showing both mongolism and the Klinefelter syndrome. *Lancet* **1**:709-710. 1959.
30. ______, K. W. Jones, P. E. Polani, J. C. deAlmedia, and J. H. Briggs. A sex-chromosome anomaly in a case of gonadal dysgenesis (Turner's syndrome). *Lancet* **1**:711-713. 1959.
31. ______, P. E. Polani, J. H. Briggs, and P. M. F. Bishop. A presumptive human XXY/XX mosaic. *Nature* **183**:1030-1032. 1959.
32. Forssman, H. and G. Hambert. Incidence of Klinefelter's syndrome among mental patients. *Lancet* **1**:1327. 1963.
33. Fraccaro, M., K. Kaijser, and J. Lindsten. Chromosome complement in gonadal dysgenesis (Turner's syndrome). *Lancet* **1**:886. 1959.
34. ______, M. Glen Bott, P. Davies, and W. Schutt. Mental deficiency and undescended testia in two males with XYY sex chromosomes. *Folia Hered. Pathol.* (Milan) **11**:211-220. 1962.
35. Franks, Robert C., Kenneth W. Bunting, and Eric Engel. Male pseudohermaphrodism with XYY sex chromosomes. *J. Clin. Endocr.* **27**:1623-1627. 1967.
36. Fraser, J. H., J. Campbell, R. C. MacGillivray, E. Boyd, and B. Lennox. The XXX syndrome—frequency among mental defectives and fertility. *Lancet* **2**:626-627. 1960.
37. Gates, William H. A case of non-disjunction in the mouse. *Genetics* **12**:295-306. 1927.
38. Hamerton, J. L. Sex chromatin and human chromosomes. *Intern. Rev. Cytol.* **12**:1-68. 1961.
39. Hauschka, Theodore S., John E. Hasson, Milton N. Goldstein,

George F. Koepf, and Avery A. Sandberg. An XYY man with progeny indicating familial tendency to non-disjunction. *Am. J. Human Genet.* **14**:22-30. 1962.

40. Hayward, M. D. and B. D. Bower. Chromosomal trisomy associated with the Sturge-Weber syndrome. *Lancet* **2**:844-846. 1960.

41. Hunter, H. Chromatin-positive and XYY boys in approved schools. *Lancet* **1**:816. 1968.

42. Hustinx, T. W. J. and A. H. F. van Olphen. An XYY chromosome pattern in a boy with Marfan's syndrome. *Genetica* **34**:262. 1963.

43. Ismail, A. A. A., R. A. Harkness, K. E. Kirkham, J. A. Loraine, P. B. Whatmore, and R. P. Brittain. Effect of abnormal sex-chromosome complements on urinary testosterone levels. *Lancet* **1**:220-222. 1968.

44. Jacobs, P. A., A. G. Baikie, W. M. Court Brown, and J. A. Strong. The somatic chromosomes in mongolism. *Lancet* **1**:710. 1959.

45. ______ and A. J. Keay. Chromosomes in a child with Bonnevie-Ullrich syndrome. *Lancet* **2**:732. 1959.

46. ______ and J. A. Strong. A case of human intersexuality having a possible XXY sex-determining mechanism. *Nature* **182**:302-303. 1959.

47. ______, A. G. Baikie, W. M. Court Brown, T. N. MacGregor, N. MacLean, and D. G. Harnden. Evidence for the existence of the human "super female." *Lancet* **2**:423-425. 1959.

48. ______, Muriel Brunton, Marie M. Melville, R. P. Brittain, and W. F. McClemont. Aggressive behaviour, mental subnormality and the XYY male. *Nature* **208**:1351-1352. 1965.

49. ______, W. H. Price, W. M. Court Brown, R. P. Brittain, and P. B. Whatmore. Chromosome studies on men in a maximum security hospital. *Ann. Human Genet.* **31**:330-347. 1968.

50. Kesaree, Nirmala and Paul V. Woolley. A phenotypic female with 49 chromosomes, presumably XXXXX. *J. Pediat.* **63**:1099-1103. 1963.

51. Kessler, Seymour and Rudolph H. Moos. XYY chromosome: premature conclusions. *Science* **165**:442. 1969.

52. Kosenow, W. and R. A. Pfeiffer. YY syndrome with multiple malformations. *Lancet* **1:**1375-1376. 1966.

53. Leff, J. P. and P. D. Scott. XYY and intelligence. *Lancet* **1:**645. 1968.

54. Lejeune, J., M. Gautier, and R. Turpin. Etude des chromosomes somatiques de neuf enfants mongoliens. *Compt. Rend. Acad. Sci.* **248:**1721-1722. 1959.

55. ———, J. Lafourcade, R. Berger and M. O. Rethore. Maladie du cri du chat et sa reciproque. *Ann. Genet.* **8:**11-15. 1965.

56. Lubs, H. A., Jr., E. V. Koenig, and L. H. Brandt. Trisomy 13-15: A clinical syndrome. *Lancet* **2:**1001-1002. 1961.

57. Lyon, M. F. Gene action in the X-chromosome of the Mouse (*Mus musculus* L.). *Nature* **190:**372-373. 1961.

58. Maclean, N., D. G. Harnden, W. M. Court Brown, Jane Bond, and D. J. Mantle. Sex-chromosome abnormalities in newborn babies. *Lancet* **1:**286-290. 1964.

59. McWhirter, Kennedy. XYY chromosome and criminal acts. *Science* **164:**1117. 1969.

59a. Melnyk, John, Frank Vanasek, Havelock Thompson, and Alfred J. Rucci. Failure of transmission of supernumerary Y chromosomes in man. Abst. Am. Soc. Human Genet. Annual Meeting, Oct. 1-4, 1969.

60. Mickey, George H. Chromosome studies in testicular feminization syndrome in human male pseudohermaphrodites. *Mammalian Chromosome Newsletter* No. 9, p. 60. 1963.

61. Migeon, Barbara R. G trisomy in an XYY male. *Human Chromosome Newsletter* No. 17. Dec. 1965.

62. Milcu, M., I. Nigoescu, C. Maximilian, M. Garoiu, M. Augustin, and Ileana Iliescu. Baiat cu hipospadias si cariotip XYY. *Studio si Cercetari de Endocrinologie* (Bucharest) **15:**347-349. 1964.

63. Miller, Orlando J. The sex chromosome anomalies. *Am. J. Obstet. Gynec.* **90:**1078-1139. 1964.

64. Minckler, Leon S. Chromosomes of criminals. *Science* **163:**1145. 1969.

65. Montagu, Ashley. Chromosomes and crime. *Psychology Today* **2:** 43-49. 1968.

66. Muldal, S. and C. H. Ockey. The "double male": a new chromosome constitution in Klinefelter's syndrome. *Lancet* **2:**492-493. 1960.

67. Nowell, P. C. and D. A. Hungerford. A minute chromosome in human granulocytic leukemia. *Science* **132:**1497. 1960.

68. Painter, T. S. The chromosome constitution of Gates "non-disjunction" (v-o) mice. *Genetics* **12:**379-392. 1927.

68a. Palmer, Catherine G. and Sandra Funderburk. Secondary constrictions in human chromosomes. *Cytogenetics* **4:**261-276. 1965.

69. Patau, K. The identification of individual chromosomes, especially in man. *Am. J. Human Genet.* **12:**250-276. 1960.

70. ______, D. W. Smith, E. Therman, S. L. Inhorn, and H. P. Wagner. Multiple congenital anomalies caused by an extra chromosome. *Lancet* **1:**790-793. 1960.

71. Penrose, L. S. The Biology of Mental Defect. Grune and Stratton, New York. 1949.

72. Pergament, Eugene, Hideo Sato, Stanley Berlow, and Richard Mintzer. YY syndrome in an American negro. *Lancet* **2:**281. 1968.

73. Pfeiffer, R. A. Der Phanotyp der Chromosomen-aberration XYY. *Wochenschrift* **91:**1355-1356. 1966.

74. Price, W. H. and P. B. Whatmore. Behaviour disorders and the pattern of crime among XYY males identified at a maximum security hospital. *Brit. Med. J.* **1:**533. 1967.

75. ______, J. A. Strong, P. B. Whatmore, and W. F. McClement. Criminal patients with XYY sex-chromosome complement. *Lancet* **1:**565-566. 1966.

76. Puck, T. T., A. Robinson, and J. H. Tjio. A familial primary amenorrhea due to testicular feminization. A human gene affecting sex differentiation. *Proc. Exper. Biol. Med.* **103:**192-196. 1960.

77. Reitalu, Juhan. Chromosome studies in connection with sex chromosomal deviations in man. *Hereditas* **59:**1-48. 1968.

78. Ricci, N. and P. Malacarne. An XYY human male. *Lancet* **1**:721. 1964.

79. Sandberg, A. A., G. F. Koepf, T. Ishihara, and T. S. Hauschka. An XYY human male. *Lancet* **2**:488-489. 1961.

80. ______, Takaaki Ishihara, Lois H. Crosswhite, and George F. Koepf. XYY genotype. *New England J. Med.* **268**:585-589. 1963.

81. Sergovich, F., G. H. Valentine, A. T. L. Chem, R. A. H. Kinch, and M. S. Smout. Chromosome aberrations in 2159 consecutive newborn babies. *New England J. Med.* **280**:851-855. 1969.

82. Smith, D. W., K. Patau, and E. Therman. The 18 trisomy syndrome and the D_1 trisomy syndrome. *Am. J. Dis. Child.* **102**:587. 1961.

83. ______, K. Patau, E. Therman, S. L. Inhorn, and R. I. Demars. The D_1 trisomy syndrome. *J. Pediat.* **62**:326-341. 1963.

84. Sohval, Arthur R. Sex chromatin, chromosomes and male infertility. *Fertility and Sterility* **14**:180-207. 1963.

85. ______. Chromosomes and sex chromatin in normal and anomalous sexual development. *Physiol. Rev.* **43**:306-356. 1963.

86. Telfer, Mary A., David Baker, Gerald R. Clark, and Claude E. Richardson. Incidence of gross chromosomal errors among tall criminal American males. *Science* **159**:1249-1250. 1968.

87. Thorburn, Marigold J., Winston Chutkan, Rolf Richards, and Ruth Bell. XYY sex chromosomes in a Jamaican with orthopaedic abnormalities. *J. Med. Genet.* **5**:215-219. 1968.

88. Tjio, J. H. and A. Levan. The chromosome number of man. *Hereditas* **42**:1-6. 1956.

89. ______, T. T. Puck, and A. Robinson. The somatic chromosomal constitution of some human subjects with genetic defects. *Proc. Natl. Acad. Sci.* **45**:1008-1016. 1959.

90. Townes, Philip L., Nancy A. Ziegler, and Linda W. Lenhard. A patient with 48 chromosomes (XYYY). *Lancet* **1**:1041-1043. 1965.

91. Vignetti, P., L. Capotorti, and E. Ferrante. XYY chromosomal constitution with genital abnormality. *Lancet* **2**:588-589. 1964.

92. Welch, J. P. D. S. Borgaonkar, and H. M. Herr. Psychopathy, mental deficiency, aggressiveness and the XYY syndrome. *Nature* **214**:500-501. 1967.

93. Wiener, Saul, and Grant Sutherland. A normal XYY man. *Lancet* **2**:1352. 1968.

D.
Genetics and Early Human Development

In the past few years, much has been written on the developmental period of individuals from fertilization to the period of maturation. Some of the deepest social and political controversies on radiation, intelligence, prenatal influences, and postnatal influences relate to material found in this section. It is important for the student to understand how researchers obtain data on these great issues.

The three papers included here show how researchers deal with three phases of early human development.

Miller's paper is a summary report "to put into perspective the major findings" of the Atomic Bomb Casualty Commission which studied the effects of exposure to atom bombs in Hiroshima and Nagasaki. There is an increased frequency of complex chromosomal aberrations among those F_1 individuals exposed while in utero when their mothers received a dose of at least 100 rad from the bomb blasts. It is evident, therefore, that there are indeed delayed and latent effects of radiation.

The paper by Record, McKeown, and Edwards deals with important variables: measured intelligence, twins, single births, maternal age, birth rank, sibling relationships, and duration of gestation. The authors suggest that twins obtain lower verbal reasoning scores (measured intelligence) than single births. This, they suggest, is due to a postnatal rather than a prenatal influence.

Eckland's paper (slightly abridged*) ranges over many areas and could have been easily placed in the first section as a significant overview

*One section in the Eckland paper has been removed from the original. The request to shorten it was mine and the choice of material to be omitted was the author's. — J. B. B.

or in the last section summing up many issues in socio-genetics. Nevertheless, it seems most appropriate to place it in this section on human development because it contains many excellent comments on intelligence and the testing of intelligence.

Eckland says neither genetics with a strict hereditary view, nor sociology with a strict environmental one, is sufficient to explain the salient problems of human development. In the section entitled "Why change" Eckland reviews the results of I. Q. performance tests with genetic relationships involving parent and child and sib and sib. (You might refer back to the paper by Abe in which identical twins were generally but not always alike in their behavioral indices.)

Throughout the paper Eckland discusses assortative mating. In its simple form assortative mating represents a nonrandom mating system whereby one chooses his or her spouse. Generally, there is a tendency for like (positive) individuals to be attracted. Eckland reports high correlations between spouses in terms of their measured intelligence and various socio-economic characteristics. Assortative mating and what this means to meritocracy is used to a great extent in the paper by Herrnstein (1971). In connection with assortative mating, it would be well to refer back to Turner's short paper on PTC tasting.

Additional Readings

Record, McKeown, and Edwards, 1969. *Annals of Human Genetics,* **33** (a) 61-69, The relation of measured intelligence to birth order and maternal age and (b) 72-79, The relation of measured intelligence to birth weight and duration of gestation. The authors deal with other factors which affect intelligence.

Beals, Anderson, and Eckland, 1967. *American Sociological Review,* **32**: 996-1001. See these three short communications for a further exchange of views on Eckland's paper.

Kato, H., 1971. Mortality in children exposed to the A-Bombs while in utero 1945-1969. *American Journal of Epidemiology,* **93**:435-442. He finds an increase of mortality in the first year of life, no increase for the next nine, and mortality again increasing after 10 years of age for those who were in utero.

Cohen, Joel, 1971. Legal abortions, socio-economic status, and measured intelligence in the United States. *Social Biology,* **18**:55-63. On preliminary data, it would appear that induced abortions could increase the apparent selection for measured intelligence. The major importance of the study shows that social actions bring about genetic consequences.

Hsia, David Yi-Yung, 1968. *Human Developmental Genetics.* Year Book Medical Publishers, Inc., Chicago. An excellent text which contains an especially pertinent section on "Genetics of prenatal growth."

Manosevitz, Martin, Gardner Lindzey, and Delbert D. Thiessen, 1969. *Behavioral Genetics: Method and Research.* An excellent collection of papers on behavioral genetics. Those intending to study psychology will find this very useful.

Kaelber, Charles T. and Thomas Pugh, 1969. Influence of uterine relations on the intelligence of twins. *New England Journal of Medicine,* **280**:1030-1034. The heavier twin of like-sex pairs was shown to have higher I.Q.s. Unequal intra-uterine relations, perhaps of a circulatory nature, are proposed as an explanation for the I. Q. differences.

Scarr, Sandra, 1969. Effects of birth weight on later intelligence. *Social Biology,* **16**.249-256. Generally children with higher birth weight were shown to have higher I. Q. scores. This paper and the one by Kaelber and Pugh come to the same conclusion in reports published in the same year.

Questions

The papers by Record and his colleagues are retrospective studies of events which have taken place. If you were to prepare a large scale prospective study dealing with the many variables they have considered, how would you approach the problem?

It has been observed that there is an increase in twins and multiple births because of the use of so-called fertility drugs. Should this have a measurable effect on the world intelligence level?

Do you know of other scientific work where two or more independent investigators arrived at essentially the same results at the same time without prior collaboration?

How would you relate the Miller paper to the previous Heller paper?

How would you relate the previous paper by Turner to the present paper by Eckland?

8. Delayed Radiation Effects in Atomic-Bomb Survivors

ROBERT W. MILLER

The pace at which radiation effects in the Japanese survivors of the atomic bombs are being reported has recently quickened. In this article I seek to put into perspective the major findings of the Atomic Bomb Casualty Commission (ABCC).

Immediately after World War II, a joint commission of the U.S. Army and Navy made observations concerning the immediate effects of exposure to the atomic bombs in Hiroshima and Nagasaki. Upon completion of its work, the joint commission recommended that the National Academy of Sciences-National Research Council conduct a study of the long-range biomedical effects of the exposures. The Council convened an advisory group, whose study of the situation in Japan led to a Presidential directive authorizing the National Research Council (NRC) to establish an organization to evaluate the delayed effects of exposure to the bombs. Thus the Atomic Bomb Casualty Commission came into existence (*1*). Its large-scale study, begun in 1948, is a cooperative venture between the NRC, representing the United States, and the Japanese National Institute of Health. The Commission's present staff of 725 Japanese and 36 foreign nationals, including 18 U.S. professionals, is collecting and analyzing data from periodic comprehensive medical examinations, from postmortem findings, and from a review of vital certificates as they are generated.

GENETIC EFFECTS

It is commonly thought that congenital anomalies are the only measure of genetic effects among children conceived after one or both of the parents have been exposed to ionizing radiation. The studies conducted at the ABCC, however, concerned six indicators of genetic damage in the F_1 generation.

Pregnancies were ingeniously ascertained (*1*). In postwar Japan, when food was in short supply, pregnant women were allowed an extra ration of rice, beginning in the fifth month of pregnancy. When such

Reprinted from *Science,* 1969, 166, 569-574 with permission of the Journal and the author.

women registered for this supplement in Hiroshima or Nagasaki, they were entered in the ABCC genetic study. From 1948 to 1953, 71,280 pregnancies were ascertained in this way, 93 percent of all that went to term in that interval. Midwives notified ABCC of each delivery they attended, and the newborns were examined in their homes by ABCC staff physicians. About 40 percent of the children were reexamined at the clinics when they were 8 to 10 months old. The results were distributed according to five levels of radiation exposure for each parent. No influence of radiation was demonstrable in this study, which, statistical tests have shown, was likely to detect a 2-fold increase in rates of malformation or a 1.8-fold increase in rates of stillbirths and deaths of newborns. Moreover, there was no effect on birth weight or on anthropometric values at 8 to 10 months attributable to radiation exposure. The sex ratio (the proportion of males to females) for children conceived after exposure of one parent to radiation will, in theory, be diminished if the mother was irradiated, and increased if the father was (*2*). In a study of about 120,000 births, such shifts in the sex ratios were found to occur in the first 10 years following detonation of the atomic bombs but not thereafter (*3*). No effect has been found on the mortality of children conceived after exposure of their parents to the bombs in Hiroshima or Nagasaki (*4*). Thus, though laboratory experimentation leaves no doubt that irradiation is mutagenic, the effect could not be demonstrated in the F_1 generation studied by ABCC.

CYTOGENETIC ABNORMALITIES

In Table 1 are summarized the results obtained in a series of studies of chromosomal abnormalities, by age group, of individuals exposed to the atomic bombs in Japan. It should be noted that, among those 30 years of age or younger at the time of the bomb who received a dose of at least 200 rad, 34 percent had complex cytogenetic abnormalities 20 years later, as compared with 1 percent of the controls (*5*). The percentage of individuals so affected who were over 30 years old at the time of the bomb was almost double the percentage of the younger group, and 4 of the 77 individuals in the older group had clones of cytogenetically abnormal cells (*6*). The frequency of complex cytogenetic abnormalities apparently increases naturally with age, from 1 percent in the younger group of the controls to 16 percent in the older group (Table 1). The complex chromosomal aberrations which occurred with increasing frequency among persons who had been over 30 years old at the time of the bomb consisted of translocations, pericentric inversions, deletions, chromatid exchanges, and centromere breaks.

Table 1. Frequencies, by age group, of complex cytogenetic abnormalities among Japanese survivors of the atomic bombs

Age group at time of the bomb	Dose (rad)	Exposed Examined (No.)	Exposed Affected (%)	Control Examined (No.)	Control Affected (%)	Reference
≤ 30 years	200+	94	34	94	1	(5)
Over 30 years	200+	77	61	80	16	(6)
In utero	100+*	38	39	48	4	(7)
Not yet conceived	150+*	103	0			(8)
Not yet conceived	100+†	25	0			(8)

*Maternal dose.
†Dose received by at least one parent.

Among persons who were *in utero* at the time of the bomb and whose mothers received a dose of at least 100 rad, 39 percent displayed complex chromosomal abnormalities as compared with 4 percent of the controls (*7*). Finding these abnormalities even among persons exposed during the first trimester *in utero* indicates that radiation can induce long-persisting changes in the lymphocyte precursors. In contrast to the cytogenetic defects observed following intrauterine (postconception) exposure, no such defects were found following preconception irradiation (exposure of either parent before conception of the child) (*8*).

These observations revealed that, in man, long-persisting chromosomal damage was induced even though an effect on the F_1 generation was not demonstrable.

EFFECTS ON THE EMBRYO

Not long after the discovery of x-rays, case reports began to appear in the medical literature describing mentally retarded children with heads of small circumference born of mothers who had received pelvic radiotherapy during early pregnancy. Fourteen of these reports were described in a publication by Murphy in 1928 (*9*). Goldstein and Murphy (*10*) identified 16 additional cases from replies to a questionnaire sent to, and completed by, a substantial number of obstetricians. In view of these findings, it is not surprising to learn that the same abnormalities were observed among children born of women who were exposed to the atomic bomb while pregnant (*11-13*). The effect was primarily among children of women who were exposed within 15 weeks of their last menstrual period (Table 2). Of the individuals examined, 56 in this category were born of mothers who had been within 1800 meters of the hypocenter. A head circumference 2 or more standard deviations below the mean for age and sex was expected (on the basis of a normal distribution) in 1.4 persons (2.5 percent) of this group but observed in 23, of whom 9 were mentally retarded. Among the 105 individuals exposed *in utero* more than 15 weeks after the mother's last menstrual period, small head circumference was expected in 2.6 but observed in 6, of whom 2 were mentally retarded. The usual frequency of comparable mental retardation among the nonexposed of the same age in Hiroshima and Nagasaki was about 1 percent (*14*). Table 3, derived from the most recent data on the group exposed in the early weeks of pregnancy (*13*), reveals a dose-response relationship; that is, the effect diminishes in frequency and severity as the distance from the hypocenter increases.

Table 2. Effects of intrauterine exposure to the Hiroshima atomic bomb [From Wood *et al.* (13)]

Gestational age (week)	Total number exposed*	Total number examined	Number with small head circumference† — Mental retardation	Number with small head circumference† — Normal intelligence
≤15	57‡	56‡	9	14
>15	109	105	2§	4‖

*Exposure within 1800 meters of the hypocenter.
†Circumference 2 or more standard deviations below average.
‡Excludes two with preexistent Down's syndrome.
§Exposed at 21 and 24 to 25 weeks, respectively.
‖Exposed at 16, 32, 32 and 36 to 40 weeks, respectively.

These findings are in accord with the results of animal experimentation and with the clinical observations by Goldstein and Murphy cited above (*10*). The malformation occurred excessively only in association with high radiation dosage and not in association with the more extensive areas of heavy destruction and economic loss, which extended far beyond the high-dosage area. Estimates of the teratogenic dose range in man

Table 3. Radiation effect on head circumference and intelligence following intrauterine exposure to the Hiroshima atomic bomb within 15 weeks of the mother's last menstrual period – S.D., standard deviation [From Wood *et al.* (13)]

	Retarded (No.)		Normal (No.)		
Distance from hypocenter (meter)	Head circumference >3 S.D. below mean	Head circumference –2 to –3 S.D.	Head circumference >3 S.D. below mean	Head circumference –2 to –3 S.D.	Total exposed*
≤1200	6	2	1	1	11
1201–1500	0†	0	2	6	23†
1501–1800	0	1	0	5	23‡
1801–2200	0	0	0	0	21‡

*Some children in the study were normal with respect to both intelligence and head circumference, thus the numbers in columns 2 to 5 do not add up to the totals in column 6. †Excludes two with preexistent Down's syndrome. ‡One not examined.

have not as yet been published, but about half of the mothers of affected children reported that they had signs of severe acute radiation sickness (*12*). No other anomalies occurred excessively among the survivors (*12, 13*), although many others have been observed in animals exposed experimentally to x-irradiation (*15*).

Fetal and infant mortality following exposure to the atomic bomb was not evaluated until 6 years after the event. Among women who were pregnant when exposed within 2000 meters of the Nagasaki hypocenter and who said they had had major signs of acute radiation sickness, 43 percent reported such mortality as compared with 9 percent of pregnant women in the same distance category who had not had acute radiation sickness (*16*). The excess is highly significant statistically ($P < .001$).

GROWTH

In 1951, as part of a comprehensive medical examination, 12 anthropometric determinations were obtained on about 2400 Hiroshima children 6 to 19 years old who had been exposed to the bomb 6 years earlier, and comparison was made with an equal number who had not been exposed (*17*). About 78 percent in each group were reexamined in 1952, and 53 percent were reexamined in 1953.

Multivariate analysis revealed that as radiation exposure increased, there were small but statistically significant decreases in body measurements at all age levels, and in growth rate at post-pubertal age levels (*18*). To some extent these differences may be due to variables other than radiation exposure—for example, to economic loss due to the blast and fires.

Nagasaki adolescents who were exposed *in utero* to radiation from the atomic bomb have been studied with respect to the mean values for several anthropometric variables (*19*). The sample of heavily exposed subjects was small. Only 16 boys and 15 girls were estimated to have received doses of 50 rad or more, and only 9 of these were exposed in the first trimester of pregnancy. Some significant differences were found which were consistent with a radiation effect.

EYE FINDINGS

In 1963 and 1964, ophthalmologic examinations were made on 1627 residents of Hiroshima and 841 residents of Nagasaki, of whom an estimated 40 percent had received doses in excess of 200 rad (*20*). In the high-dose group there were significantly more axial opacities seen by ophthalmoscope and confirmed by slit-lamp examination than in groups more distantly exposed. Most of these lesions were small, and only one was regarded as a mature radiation cataract. Another finding, more in the nature of a measure for biologic dosimetry than anything else, was a polychromatic sheen, sometimes granular, in the posterior subcapsular area of the lens as visualized by slit-lamp biomicroscopy. A dose-response relationship was demonstrated. These abnormalities did not affect visual acuity. In previous ophthalmologic examinations by ABCC, less than a dozen survivors were classified as having severe radiation cataracts, and in none was visual acuity worse than 20/70 (*21*).

In the most recent survey (*20*), there was a suggestion of a dose-related impairment of visual acuity among children who were *in utero* at the time of exposure to the bomb, but the sample size was too small for the test to be of statistical significance. In brief, though some ophthalmic effects of irradiation have been noted among atomic-bomb survivors, impairment of vision has been rare, and relatively mild.

LEUKEMOGENESIS

The leukemogenic effect of radiation in man was suspected at about the same time that teratogenic effects were—again, from case reports (*22*). Lymphoma was experimentally induced in mice in 1930 (*23*). Then, by simply reviewing the death notices published weekly in the *Journal of the American Medical Association,* Henshaw and Hawkins (*24*) found that leukemia was reported 1.7 times more often as a cause of death among U.S. physicians, a group occupationally exposed to x-rays, than among the general population of adult white males. Using the same source, Ulrich and March independently found that U.S. radiologists died significantly more often of leukemia than other physicians did (*25*). In consequence, a leukemogenic effect of exposure to atomic radiation was expected among the survivors in Hiroshima and Nagasaki, and it was found (*26*). A dose-response relationship was observed which can be attributed to no variable except radiation. A peak in occurrence was reached in 1951, more marked for the acute leukemias than for chronic granulocytic leukemia. Chronic lymphocytic leukemia, rare among the Japanese (*27*), did not increase in frequency. In children, leukemia was generally acute, as it usually is in children, the lymphocytic form being as common as the granulocytic. In all age groups acute leukemia continued to occur at higher than usual rates through 1966, whereas chronic granulocytic leukemia had fallen to near-normal rates. The ABCC study leaves no doubt that whole-body exposure to ionizing radiation at sufficiently high doses can induce leukemia in man.

Human leukemia may also be induced by partial-body irradiation, as indicated by the dose-response effect observed in British men given radiotherapy for ankylosing spondylitis (*28*). Again, the peak was reached about 5 years after the first exposure (the first course of therapy). The predominant form in these adults was granulocytic; no increase occurred in the frequency of chronic lymphocytic leukemia. In all, 52 cases of leukemia was observed, as compared with the 5.48 expected on the basis of national mortality rates for England and Wales, and 15 persons developed aplastic anemia (perhaps subclinical leukemia) as compared with the .051 expected. This study, in conjunction with the ABCC study, revealed that ionizing radiation can induce more than one form of leukemia in man, but not all forms, the notable exception being chronic lymphocytic leukemia.

Irradiation is but one of several circumstances that carry exceptionally high risk of leukemia (*29*). At the highest risk yet known is the child whose identical twin develops leukemia before the age of 6 years. The probability is 1 in 5 that the second twin will develop the disease within weeks or months after the first child falls ill. In about the same category are the person with polycythemia vera treated with x-ray or phosphorus-

32 (or both) and persons with Bloom's syndrome or Fanconi's anemia. The probability of developing leukemia was substantially less for heavily exposed Hiroshima survivors—about 1 in 60 individuals were so affected within 12 years of exposure. At still lower risk of leukemia are children with Down's syndrome (1 in 95 for children under 10 years old) and radiation-treated patients with ankylosing spondylitis (1 in 270 were so affected within 15 years after radiotherapy). These groups are alike in that each has a distinctive genetic feature, but these features are not of a single type. Identical twins have identical genes; Bloom's and Fanconi's syndromes are heritable disorders characterized by chromosomal fragility; in radiation-treated polycythemia vera, aneuploidy has been described in a substantial proportion of cases before radiation and chromosomal breaks are regularly found following radiotherapy; atomic-bomb survivors (and persons exposed to ionizing radiation from other sources) exhibit long-lasting chromosomal breaks; and in Down's syndrome there is congenital aneuploidy.

Leukemia in patients with polycythemia vera has been attributed to radiotherapy (*30*). It should be noted, however, that the probability of occurrence of the neoplasm was 10 times as high in these patients as it was among heavily exposed survivors of the Hiroshima bomb and 45 times as high as it was in radiation-treated ankylosing spondylitis. One must conclude either that radiation exposure or damage is greater in polycythemia vera than in the other two instances or that polycythemia vera predisposes to leukemia in the absence of radiation exposure.

Several studies, considered individually, suggest that very small exposures to radiation before conception or during pregnancy may increase by 50 percent the child's risk of leukemia. When these studies are considered collectively, however, there is reason to suspect that some fault in the methods, difficult or impossible to escape, may be implicating radiation spuriously.

The individual results are as follows. In 1958, Stewart and her associates (*31*) described a study of 677 leukemic children in England and Wales in which the proportion of mothers who reported abdominal exposure to *diagnostic* radiation during the relevant pregnancy was twice the proportion for mothers of normal children living in the same area. A similar difference was reported with respect to 739 children with neoplasms of other kinds. It is possible that mothers of children with cancer reported their radiologic exposures more fully than the mothers of healthy children did. MacMahon (*32*) avoided this potential bias in histories obtained through interviews by studying obstetric records for irradiation during pregnancy among mothers of 304 leukemic children and 252 children with other cancer as compared with records for a 1-percent sample of all other births in the area (New England). He found a 40-percent excess of (i) leukemia and (ii) all other neoplasia among children whose mothers' records showed diagnostic radiation of the abdomen during pregnancy. Similar results were obtained by Graham *et al.* (*33*) with respect to leukemia, and by Stewart and Kneale (*34*) for each of six categories of childhood cancer. A causal relationship would be indicated if a dose-response effect could be demonstrated, if the results were consistent with those from animal experimentation, and if concomitant variables could be excluded. The exposures involved were too small to permit evaluation of a dose-response effect, there are no data from animal studies which support the observations in man, and the condition being treated by the radiologic procedure, rather than the x-ray exposure itself, could, in theory at least, be the oncogenic factor.

Recently Graham and his associates (*33*) described an excess of leukemia among children whose mothers *or* fathers gave histories of diagnostic radiation exposure up to a decade before the children were conceived. Again, there are no animal studies to support this observation. Moreover, in a prospective study (*35*) of 22,400 children conceived after their parents had been heavily exposed to radiation from the atomic bombs in Hiroshima or Nagasaki, no excess of leukemia was found.

Table 4 summarizes the results following very small doses of x-ray, and indicates that such irradiation was equally oncogenic whether exposure occurred before conception or during pregnancy, whether the neoplasm studied was leukemia or any other major cancer of childhood, and whether the study was based on interviews, which may be biased, or on hospital records. Taken in the aggregate, the similarity of results in the absence of a dose-response effect or of supporting data from animal experimentation raises a question about the biologic plausibility of a causal relationship. In particular one must ask, in the absence of demonstrable mutagenic or cytogenetic abnormalities in the F_1 generation, if irradiation of the parent prior to conception is likely to induce leukemia in the child.

OTHER CANCER

Wanebo *et al.* (*36*) have recently reported that "accumulated information . . . strongly suggests that exposure to ionizing radiation has in-

Table 4. Relative risk of various childhood cancers following intrauterine or preconception exposures to diagnostic radiation

Neoplasm	Relative risk*
Intrauterine exposure	
Stewart and Kneale (33)	1.5
Leukemia	1.5
Lymphosarcoma	1.5
Cerebral tumors	1.5
Neuroblastoma	1.5
Wilms' tumor	1.6
Other cancer	1.5
MacMahon (31)	
Leukemia	1.5
Central-nervous system tumors	1.6
Other cancer	1.4
Graham *et al.* (32)	
Leukemia	1.4
Preconception exposure	
Graham *et al.* (32)	
Leukemia	
Mother exposed	1.6
Father exposed	1.3

*Relative risk in controls = 1.0.

creased the risk of lung cancer among atomic bomb survivors." These investigators observed 17 such cases, as compared with 9 expected (dose, 90 rad or more). A weakness in the report was the finding that the lung cancers induced were nonspecific as to histologic type, rather than of the undifferentiated or small-cell type, as in U.S. uranium miners and in workers heavily exposed to mustard gas, a radiomimetic chemical (*37*).

Wanebo *et al.* (*38*) have reported that "information on breast cancer among survivors of the atomic bombings of Hiroshima and Nagasaki has now accumulated to the point where a fairly definite carcinogenic effect seems established." Six cases were observed among women who were exposed to 90 rad or more, as compared with 1.53 cases expected—an excess of only 4.5 cases. There was no specificity as to histologic type.

It may be difficult or impossible to avoid certain biases that could produce such a small excess—for example, unequal detection of cases with respect to exposure category, or dissimilar cancer risks in relation to some variable other than radiation which distinguished the heavily exposed from others in the study. Wanebo *et al.* considered the possibility of biases and believed that none were present. The absence of a dose-response relationship makes interpretation of the results difficult, as does the small or uncertain effect observed in studies of other exposed persons.

Wood *et al.* (*39*) have recently described an excess of thyroid cancer among Japanese survivors of the atomic bomb. The increase was greater in women than in men, the effect being proportionate to the radiation dose, but no specificity as to cell type was found. These observations are in accord with the results of animal experimentation and with the increase in frequency of thyroid cancer following therapeutic radiation early in childhood (*40*).

One may conclude that, among the Japanese survivors of the atomic bomb, only leukemia and thyroid cancer have been shown to be radiation-induced. The evidence pertaining to cancer of the breast or lung is still very much in doubt.

MORTALITY

Animal experimentation has shown that ionizing radiation can induce a shortening of life span which is attributable to no specific disease but to an accelerated occurrence of disease in general (*41*). The ABCC has conducted a study of life span among the survivors of the atomic bombs. The most recent published analysis concerns deaths in Hiroshima and Nagasaki, in the decade 1950 to 1960, in a sample of 99,393 persons—survivors of all ages—and a similar group of individuals not exposed to the bomb (*42*). When leukemia was excluded as a cause of death, the mortality ratios for exposed persons who had been within 1200 meters of the hypocenter in Hiroshima and Nagasaki were elevated by about 15 percent, an increase that was statistically significant when the data for both sexes and both cities were evaluated through a combined test. The increase was greater for women than for men and faded with time, reaching near-normal rates in about 1955. In another analysis of mortality, now in progress, the data through 1966 are being evaluated to determine if, after an extended period of latency, mortality may again be increased.

SUMMARY

Since 1948 the ABCC has been evaluating the health of survivors of the atomic bombs in Hiroshima and Nagasaki. In a study of about 70,000 children conceived after the explosion, six indicators of genetic damage failed to reveal an unequivocal effect of radiation. Furthermore, this group displayed no evidence of cytogenetic abnormality, in contrast to the increased frequency of complex chromosomal aberrations found among those exposed *in utero* or at any time during the entire life span. The effect was most pronounced among persons whose exposures occurred when they were 30 years of age or older.

Although a wide variety of congenital malformations have been produced in experimental animals by irradiation of the pregnant mother, the only anomaly observed among the Japanese survivors to date has been small head circumference associated with mental retardation, the effect being proportionate to the radiation dose.

The ABCC study leaves no doubt that whole-body irradiation in sufficient dose is leukemogenic in man. A similar effect following partial-body irradiation has been observed among British men given radiotherapy for ankylosing spondylitis. In both studies the effect was proportionate to the dose, the peak occurred about 6 years after first exposures, and the increase was in acute leukemias and chronic granulocytic leukemia, not in the chronic lymphocytic form of the disease.

In the past few years, a high risk of leukemia has been associated with several human attributes and with radiation exposure. These circumstances have in common an unusual genetic feature, though not of a single type.

In several studies conducted in the United States or Great Britain, very small doses of x-ray were reported to be equally oncogenic whether exposure occurred before conception or during intrauterine life; whether the neoplasm studied was leukemia or any other major cancer of childhood; and whether the study was based on interviews, which are subjective, or on hospital records, which are not. Among the features that argue against a causal relationship are the similarity of results despite the dissimilarity of subject matter and, with regard to radiation before the child's conception, the failure, in a prospective study by ABCC, to find an excess of leukemia in 22,400 children conceived after their parents had been heavily exposed to radiation from the atomic bomb.

Increases in cancers other than leukemia have recently been reported among the Japanese survivors. Twice the normal frequency of lung cancer was found among persons exposed to doses of 90 rad or more, in a study handicapped by failure to demonstrate specificity with regard to histologic type, as in U.S. uranium miners. A report of an excess of breast cancer was based on 6 cases observed as compared with 1.53 expected among women who were exposed to doses of 90 rad or more. Certain biases, difficult or impossible to avoid, could produce this small excess. Thyroid cancer, on the other hand, does appear to have been induced by radiation, since a dose-response relationship was apparent and the results are consistent with those observed following therapeutic irradiation.

Other effects attributable to radiation but relatively small in magnitude were an increase in general mortality, exclusive of death from leukemia, during the first 10 years after exposure; a statistically significant but biologically small retardation in growth and development; infrequent radiation cataracts, none of which greatly diminished visual acuity; and a polychromatic sheen on the posterior subcapsule of the lens of the eye, which caused no disability but was related to radiation dose.

References and Notes

1. J. V. Neel and W. J. Schull, "The Effect of Exposure to the Atomic Bombs on Pregnancy Termination in Hiroshima and Nagasaki," *Nat. Acad. Sci. Nat. Res. Counc. Publ. No. 461* (1956).
2. J. V. Neel, *Changing Perspectives on the Genetic Effects of Radiation* (Thomas, Springfield, Ill., 1963).
3. W. J. Schull, J. V. Neel, A. Hashizume, *Amer. J. Hum. Genet.* **18,** 328 (1966).
4. H. Kato, W. J. Schull, J. V. Neel, *ibid.,* p. 339.
5. A. D. Bloom, S. Neriishi, N. Kamada, T. Iseki, R. J. Keehn, *Lancet* 1966-**II,** 672 (1966).
6. A. D. Bloom, S. Neriishi, A. A. Awa, T. Honda, P. G. Archer, *ibid.* 1967-**II,** 802 (1967).

7. A. D. Bloom, S. Neriishi, P. G. Archer, *ibid.* 1968-**II,** 10 (1968).
8. A. A. Awa, A. D. Bloom, M. C. Yoshida, S. Neriishi, P. G. Archer, *Nature* **218,** 367 (1968).
9. D. P. Murphy, *Surg. Gynecol. Obstet. Int. Abstr. Surg.* **47,** 201 (1928).
10. L. Goldstein and D. P. Murphy, *Amer. J. Roentgenol. Radium Ther. Nucl. Med.* **22,** 322 (1929).
11. G. Plummer, *Pediatrics* **10,** 687 (1952).
12. R. W. Miller, *ibid.* **18,** 1 (1956).
13. J. W. Wood, K. G. Johnson, Y. Omori, *ibid.* **39,** 385 (1967).
14. J. W. Wood, K. G. Johnson, Y. Omori, S. Kawamoto, R. J. Keehn, *Amer. J. Public Health Nat. Health* **57,** 1381 (1967).
15. R. Rugh, *Ann. Rev. Nucl. Sci.* **9,** 493 (1959).
16. J. N. Yamazaki, S. W. Wright, P. M. Wright, *Amer. J. Dis. Child.* **87,** 448 (1954).
17. E. L. Reynolds, *Atomic Bomb Casualty Comm. Tech. Rep.* (1954), pp. 20-59.
18. J. V. Nehemias, *Health Phys.* **8,** 165 (1962).
19. G. N. Burrow, H. B. Hamilton, Z. Hrubec, *J. Amer. Med. Ass.* **192,** 97 (1965).
20. M. D. Nefzger, R. J. Miller, T. Fujino, *Amer. J. Epidemiol.* **89,** 129 (1968).
21. D. G. Cogan, S. F. Martin, H. Ikui, *Trans. Amer. Ophthalmol. Soc.* **48,** 62 (1950); R. J. Miller, T. Fujino, M. D. Nefzger, *Arch. Ophthalmol.* **78,** 697 (1967).
22. C. E. Dunlap, *Arch. Pathol.* **34,** 562 (1942).
23. C. Krebs, H. C. Rask-Nielsen, A. Wagner, *Acta Radiol. Suppl.* **10,** 1 (1930).
24. P. S. Henshaw and J. W. Hawkins, *J. Nat. Cancer Inst.* **4,** 339 (1944).
25. R. C. March, *Radiology* **43,** 276 (1944); H. Ulrich, *N. Engl. J. Med.* **234,** 45 (1946).
26. J. H. Folley, W. Borges, T. Yamawaki, *Amer. J. Med.* **13,** 311 (1952); A. B. Brill, M. Tomonaga, R. M. Heyssel, *Ann. Intern.*

Med. **56,** 590 (1962); O. J. Bizzozero, Jr., K. G. Johnson, A. Ciocco, *N. Engl. J. Med.* **274,** 1095 (1966).

27. S. C. Finch, T. Hoshino, T. Itoga, M. Ichimaru, R. H. Ingram, Jr., *Blood* **33,** 79 (1969).
28. W. M. Court Brown and R. Doll, *Leukaemia and Aplastic Anaemia in Patients Irradiated for Ankylosing Spondylitis* (Her Majesty's Stationery Office, London, 1957); *Brit. Med. J.* 1965-**II,** 1327 (1965).
29. R. W. Miller, *Cancer Res.* **27,** 2420 (1967).
30. B. Modan and A. M. Lilienfeld, *Medicine* **44,** 305 (1965).
31. A. Stewart, J. Webb, D. Hewitt, *Brit. Med. J.* 1958-**I,** 1495 (1958).
32. B. MacMahon, *J. Nat. Cancer Inst.* **28,** 1173 (1962).
33. S. Graham, M. L. Levin, A. M. Lilienfeld, L. M. Schuman, R. Gibson, J. E. Dowd, L. Hempelmann, *Nat. Cancer Inst. Monogr.* **19,** 347 (1966). Subsequent reanalysis of the data for children under 4 years old suggested that cofactors are involved—for example, maternal history of fetal mortality, or virus infection in the child more than 12 months before the diagnosis of leukemia [R. W. Gibson, I. D. J. Bross, S. Graham, A. M. Lilienfeld, L. M. Schuman, M. L. Levin, J. E. Dowd, *N. Eng. J. Med.* **279,** 906 (1968)].
34. A. Stewart and G. W. Kneale, *Lancet* 1968-**I,** 104 (1968).
35. T. Hoshino, H. Kato, S. C. Finch, Z. Hrubec, *Blood* **30,** 719 (1967).
36. C. K. Wanebo, K. G. Johnson, K. Sato, T. W. Thorslund, *Amer. Rev. Resp. Dis.* **98,** 778 (1968).
37. J. K. Wagoner, V. E. Archer, F. E. Lundin, Jr., D. A. Holaday, J. W. Lloyd, *N. Eng. J. Med.* **273,** 181 (1965); S. Wada, Y. Nishimoto, M. Miyanishi, S. Kambe, R. W. Miller, *Lancet* 1968-**I,** 1161 (1968).
38. C. K. Wanebo, K. G. Johnson, K. Sato, T. W. Thorslund, *N. Engl. J. Med.* **279,** 667 (1968).
39. J. W. Wood, H. Tamagaki, S. Neriishi, T. Sato, W. F. Sheldon, P. G. Archer, H. B. Hamilton, K. G. Johnson, *Amer. J. Epidemiol.* **89,** 4 (1969).
40. S. Lindsey and I. L. Cheikoff, *Cancer Res.* **24,** 1099 (1964); L. H. Hempelmann, *Science* **160,** 159 (1968).

41. J. B. Storer, *Radiation Res.* **25,** 435 (1965).
42. S. Jablon, M. Ishida, M. Yamasaki, *ibid.,* p. 25.

9. An Investigation of the Difference in Measured Intelligence between Twins and Single Births

R. G. RECORD, THOMAS McKEOWN, AND J. H. EDWARDS

Several investigations have shown that mean scores in intelligence tests are lower (by about 5 points on the conventional scale with mean of 100) for twins than for single births (Mehrotra & Maxwell, 1949; Sandon, 1957; Drillien, 1961). The reasons for the difference are by no means clear. It has been suggested on the one hand that they may result mainly from prenatal influences (Churchill, 1965) and on the other that they are more likely to be due to postnatal verbal handicaps experienced by twins (Day, 1932; Davis, 1937; McCarthy, 1954; Lewis, 1963).

The present investigation of the difference in measured intelligence between twins and singletons is based on the population of children used previously in examination of variation in verbal reasoning scores of single births (Record, McKeown & Edwards, 1969*a, b*). This population comprised all Birmingham live births in the period 1 January 1950 to 1 September 1954. Numbers of children born and numbers whose V.R. scores in the eleven-plus examination were matched were as follows:

	Number of liveborn children	Number of children whose V.R. scores were matched
Single births	84,341	48,913
Twins	2,259	1,242
Triplets	30	17
	86,630	50,172

The differences are accounted for mainly by those who left the City or died before age 11, or did not take the examination because they were

Reprinted by permission from *Annals of Human Genetics,* 1970, Volume 34, 11-20. This research was assisted by a grant from the Association for the Aid of Crippled Children, New York, and by a Research Grant from the Medical Research Council to one of us (T. McK.).

ineducable, in private schools, or in special schools for the handicapped, or though in ordinary schools, had been assessed as 'borderline subnormal'. As would be expected, the proportion of liveborn children who did not take the examination was somewhat higher for multiple than for single births.

The linkage of birth data to examination results was effected by a computer working to rather rigid standards of acceptability of the identity of the two records. In order to increase the number of twins for the present analysis the unmatched twins were reviewed and by manual methods a further 167 acceptable linkages were achieved, bringing the number of twins to 1409.

However, this total was considered too small for some purposes and the series was extended by inclusion of births for the subsequent 3 years (1 September 1954 to 1 September 1957). In this period there were 1474 liveborn twins and twenty-nine liveborn triplets, of whom 755 and ten respectively took the Birmingham eleven-plus examination. The analysis is therefore based on 2164 twins and (for Table 1 only) twenty-seven triplets born in a period of approximately 8 years.

Since part of the analysis relies on a comparison with single children born in the first 5 of the 8 years, it was necessary to enquire whether standards of marking had changed in the last 3 years. Mean V.R. scores of twins examined in the two periods suggest that they did not.

	Twins born 1950–54	Twins born 1954–57
Number	1409	755
Mean V.R. scores	95.52	95.92
S.D.	14.52	14.76

The difference between mean scores is 0.40 (standard error 0.66) and between standard deviations is 0.24 (standard error 0.47). These results suggest that there was no considerable change in standards of marking between the two periods and that twins born in the 8 years 1950–57 can be compared with single children born in the first 5.

Mean V.R. scores according to number of births are given in Table 1. The difference between scores of singletons and twins (4.4) is of the same order as that reported by Mehrotra & Maxwell (1949); the difference between singletons and triplets (8.5) is about twice as great.

Table 1. Mean V.R. scores of single births, twins and triplets

	Males	Females	Total
(a) Single births	99.4	100.9	100.1
	(24,348)	(24,565)	(48,913)
(b) Twins	94.6	96.8	95.7
	(1,105)	(1,059)	(2,164)
(c) Triplets	91.0	93.5	91.6
	(27)	(6)	(33)
Difference (a)–(b)	4.8	4.1	4.4
Difference (a)–(c)	8.4	7.4	8.5

In this and other tables, numbers in brackets are the numbers of children.

Numbers of triplets are too small for detailed examination and the remainder of the analysis is concerned with investigation of the following possible influences on the singleton/twin differences: (*a*) maternal age and birth rank; (*b*) birth weight and duration of gestation; (*c*) type of twinning; (*d*) order of birth; and (*e*) survival of co-twin.

MATERNAL AGE AND BIRTH RANK

Since V.R. scores vary according to maternal age and birth rank (Illsley, 1967; Record *et al.* 1969*a*), and since the distribution of twins is also related to these variables, it is desirable at the outset to enquire to what extent the difference between scores of twins and single births can be attributed to this association.

Table 2 shows the well-known increase in frequency of twinning with increasing age and with increasing number of previous sibs. Table 3 gives scores of twins and single births according to maternal age and number of sibs. In spite of some irregularity due to small numbers, the pattern is the same for twins as for singletons: scores increase with increasing age and decrease with increasing number of sibs. In every cell, however, mean scores are lower for twins than for single births.

By standardizing the scores of the single children to the maternal age distribution of the twins the twin/singleton difference is increased and by standardizing for birth rank the difference is decreased (see Fig. 1). This result could be predicted from Tables 2 and 3. Less predictable

Table 2. Ratio of numbers of twins to numbers of single births at different maternal ages and birth ranks

Previous sibs	Maternal age: Under 25	25–	30–	35 and over
0	1.0*	1.4	1.3	1.4
	(211)	(165)	(66)	(29)
1	1.4	1.9	2.1	2.5
	(123)	(234)	(166)	(84)
2	1.9	2.2	3.0	2.5
	(50)	(144)	(184)	(90)
3 and over	1.3	1.9	3.0	2.8
	(12)	(101)	(244)	(209)

Based on 2162 twins and 41,195 single births of known maternal age and birth rank.
*The ratio number of twins/number of singletons in this cell has been taken as one; the ratios in other cells are adjusted accordingly.
Numbers of twins are given in brackets.

is the effect of standardization in respect of both variables. This reduces mean scores of singletons slightly (from 100.1 to 99.6) but leaves an appreciable difference between twins and singletons (3.9) unexplained.

Table 3. Mean V.R. scores of twins and single births according to maternal age and number of previous births

Previous sibs	Maternal age: Under 25		25–		30–		35 and over		Total	
	Twin	Single	Twin	Single	Twin	Single	Twin	Single	Twin	Single
0	95.9	100.5	100.7	104.9	99.3	106.7	104.5	107.5	98.6	102.9
1	96.3	97.5	95.8	102.0	102.4	104.1	101.2	104.8	98.5	101.6
2	89.6	94.0	93.5	97.6	98.3	100.8	97.6	101.8	95.8	98.9
3 and over	86.9	92.2	85.7	93.1	90.0	94.3	93.3	95.8	90.6	94.5
Total	94.9	99.0	95.0	100.8	96.4	100.9	96.3	100.0	95.7	100.1

Based on 2162 twins and 41,195 single births of known maternal age and birth rank.

It is concluded that only a small part of the score difference can be attributed to the increased frequency of twinning with increasing maternal age and birth rank.

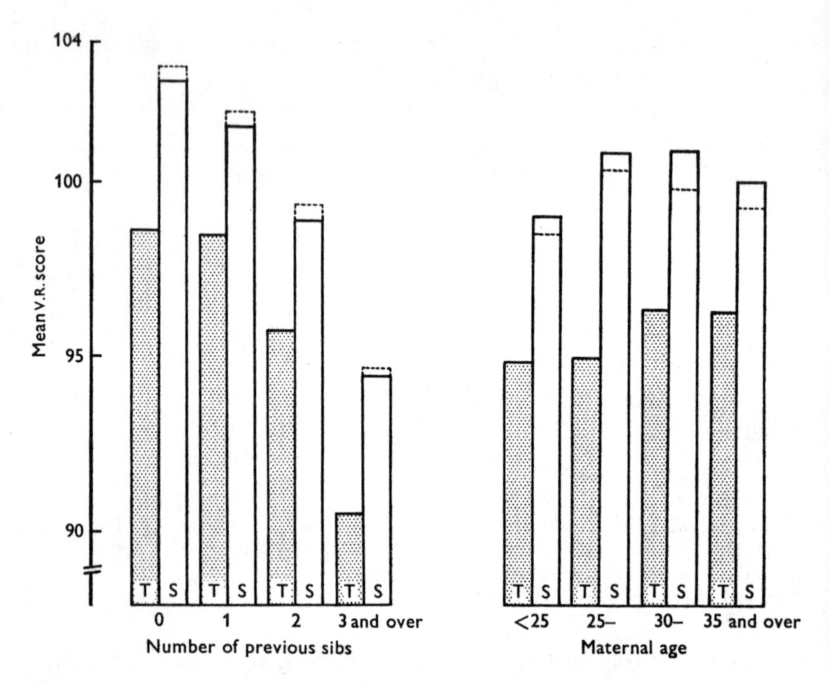

Fig. 1 Mean V.R. scores of twins (T) and single (S) children according to number of previous sibs and maternal age. Dotted lines show mean scores of single children standardized in the two cases respectively to distribution according to maternal age and number of previous sibs of twins.

BIRTH WEIGHT AND DURATION OF GESTATION

It was reported previously for single births that scores increase with increasing birth weight and that children born early or late have lower scores than those from pregnancies of average duration (Record *et al.* 1969*b*). Table 4 shows that the same relationships exist for twins, although again there is some irregularity due to small numbers. In most cells values are higher for singletons than for twins, which suggests that the total twin/singleton difference may not be explained by the distributions according to birth weight and duration of gestation.

Results of standardization for the two variables are also consistent with this conclusion. By standardization of the singletons to the twin distribution it was noted that weight differences have more effect than gestation differences; but since it is very unlikely that twins and single births of the same weight and duration of gestation are alike in all other

Table 4. Mean V.R. scores of twins and single births according to birth weight and duration of gestation

Duration of Gestation (weeks)	Birth weight (kg.)										Total	
	Less than 2.0		2.0–		2.5–		3.0–		3.5 and over			
	Twin	Single	Twin	Single	Twin	Single	Twin	Single	Twin	Single	Twin	Single
Less than 32	94.6	92.6									94.4	93.8
	(32)	(60)	(1)	(27)	(1)	(11)	(0)	(15)	(0)	(3)	(34)	(116)
32–	96.6	92.1	92.0	96.2							95.8	93.6
	(69)	(63)	(21)	(54)	(7)	(38)	(2)	(19)	(0)	(12)	(99)	(186)
34–	93.8	95.3	92.6	96.4	91.1	98.0					92.6	96.2
	(60)	(67)	(86)	(212)	(23)	(161)	(4)	(125)	(0)	(71)	(173)	(636)
36–	90.8	94.9	93.9	97.1	94.5	97.8	98.8	98.0			93.9	97.4
	(57)	(52)	(210)	(354)	(111)	(743)	(20)	(632)	(2)	(389)	(400)	(2,170)
38–	93.6	91.7	96.6	97.2	97.2	99.4	97.5	101.0	101.8	100.9	97.0	100.3
	(32)	(49)	(197)	(541)	(254)	(2,667)	(115)	(4,310)	(18)	(2,617)	(616)	(10,184)
40–	88.8	92.0	95.1	95.4	99.1	98.6	97.9	100.5	99.2	101.9	97.2	100.7
	(26)	(23)	(91)	(454)	(148)	(3,576)	(103)	(9,062)	(22)	(9,342)	(390)	(22,457)
42 and over			95.7	94.4	93.8	97.0	95.7	98.7			95.5	99.1
	(4)	(3)	(14)	(106)	(25)	(762)	(14)	(2,119)	(6)	(2,795)	(63)	(5,785)
Total	93.3	93.3	94.8	96.3	96.8	98.6	97.4	100.3	101.2	101.2	95.7	100.1
	(280)	(317)	(620)	(1,748)	(569)	(7,958)	(258)	(16,282)	(48)	(15,229)	(1,775)	(41,534)

Means have not been calculated for cells in which the number of twins is less than 10.

respects, standardization for these variables is a procedure of doubtful validity. For this reason the results have not been given in Table 4.

A more acceptable basis for assessment of the significance of birth weight is a comparison of scores of twins from the same pairs. In Table 5 pairs with no weight difference have been excluded, along with those whose birth weights could not be related to their V.R. scores. Scores of the remaining 857 pairs are distributed according to birth weight differences between twins of the same pair. Where weight differences are small or moderate, score differences are trivial; and only when weight differences are more than two-thirds of a kilogram—a very substantial variation between twins of the same pair—is there an appreciable difference between scores.

Table 5. Mean V.R. scores of paired twins according to the difference between their birth weights

		Birth weight difference (kg.)					
Sex		Less than 0.2	0.2–	0.45–	0.7–	0.9 and over	Total
Male	Heavier	95.4	93.4	92.9	93.4	98.2	94.4
Male	Lighter	94.1	95.0	91.7	91.8	92.2	93.7
		(111)	(82)	(47)	(20)	(12)	(272)
Female	Heavier	97.6	97.0	99.7	102.3	97.1	98.1
Female	Lighter	96.9	97.1	97.7	97.9	95.2	97.1
		(93)	(70)	(50)	(14)	(13)	(240)
Male	Heavier	91.6	92.6	92.2	97.6	98.4	93.3
Female	Lighter	94.9	91.7	95.7	96.7	98.0	94.5
		(62)	(67)	(38)	(27)	(14)	(208)
Female	Heavier	98.8	96.7	99.7	102.4	103.8	98.8
Male	Lighter	95.8	94.7	97.4	96.7	98.4	95.9
		(62)	(39)	(21)	(10)	(5)	(137)
Total (giving	Heavier	95.9	94.9	96.1	98.9	99.4	96.1
equal weighting	Lighter	95.4	94.6	95.6	95.8	96.0	95.3
to each group)	Difference	0.5	0.3	0.5	3.1	3.4	0.8
		(328)	(258)	(156)	(71)	(44)	(857)

Thirty-three pairs with no weight difference have been excluded.
Seventy-three pairs could not be used for this analysis because their birth weights could not be related to their V.R. scores.

These results suggest that twin/singleton differences in V.R. scores cannot be attributed largely to their shorter duration of gestation or retarded foetal growth, except possibly where the retardation is very marked.

TYPE OF TWINNING

It is known that prenatal hazards are greater for monozygotic than for dizygotic twins: their gestation is slightly shorter; birth weight is lower (McKeown & Record, 1952); and stillbirth and neonatal mortality is considerably higher (Record, Gibson & McKeown, 1952). It is conceivable, therefore, that the risks associated with zygosity may contribute to the low twin V.R. scores; if so there would be a difference between scores of monozygotic and dizygotic twins.

Although zygosity of twins was not examined, if it were related to intelligence this would be expected to emerge in differences between scores of like-sex and unlike-sex pairs. Mean scores of paired twins, where both members survived and took the examination, were slightly higher for like-sex than for unlike-sex pairs (by 0.3 for males and by 1.2 for females: Table 6). Differences in birth weight were trivial.

Table 6. Mean V.R. scores and mean birth weight of paired twins

		Like-sex pairs (a)	Unlike sex pairs (b)	Difference (a)–(b)
Males	V.R. score	94.5	94.2	0.3
	Birth weight (kg.)	2.56	2.58	–0.02
		(618)	(358)	
Females	V.R. score	97.3	96.1	1.2
	Birth weight (kg.)	2.41	2.45	–0.04
		(590)	(358)	

Two female twins whose birth weights were not recorded have been excluded.

This evidence suggests that increased risks associated with monozygosity have little influence on V.R. scores and are therefore unlikely to contribute to the differences in measured intelligence between twins and single births.

ORDER OF BIRTH

Since perinatal mortality is considerably higher for the second twin delivered than for the first (Camilleri, 1963), it seems probable that surviving second-born twins also experience risks which might impair their subsequent development. If so, V.R. scores might be lower for the second than for the first twin delivered.

Table 7. Mean V.R. scores and mean birth weight of paired twins according to order of birth

		V.R. score			Birth weight (kg.)		
		First twin (a)	Second twin (b)	Difference (a)–(b)	First twin (a)	Second twin (b)	Difference (a)–(b)
Like-sex pairs	Males	94.1 (270)	94.4 (270)	–0.3	2.60 (270)	2.54 (270)	0.06
	Females	97.3 (239)	97.9 (239)	–0.6	2.43 (239)	2.39 (239)	0.04
Unlike-sex pairs	Males	94.5 (160)	94.1 (182)	0.4	2.69 (160)	2.59 (182)	0.10
	Females	96.4 (182)	95.9 (160)	0.5	2.50 (182)	2.52 (160)	–0.02
Total (giving equal weighting to each group)		95.6 (851)	95.6 (851)	0.0	2.55 (851)	2.51 (851)	0.04

224 twins whose birth order was not recorded have been excluded.

However, Table 7 provides no support for this conclusion. Among pairs from which both twins survived, mean scores for first born are slightly higher than for second born in unlike-sex pairs and slightly lower in like-sex pairs; when both types are considered together there is no difference. Differences in birth weight between first- and second-born twins are also very small. On this evidence, the considerable risks to which the second twin is exposed during birth do not contribute to the low V.R. scores of twins.

Although they have no direct bearing on the twin/singleton scores differences, the correlation between scores of first- and second-born twins is of interest and the coefficients are given in Table 8. As would be expected, values are higher for like-sex than for unlike-sex pairs. The estimates are very similar to those reported previously by Mehrotra & Maxwell (1949): MM, 0.69; FF, 0.75; MF, 0.63.

Table 8. Correlation between V.R. scores of twins and co-twins

		No. of pairs	Coefficient of correlation
Like-sex twins:	both male	309	0.67
	both female	296	0.80
	total	605	0.74
Unlike-sex twins		358	0.62
All twins		963	0.70

SURVIVAL OF CO-TWIN

Willerman & Churchill (1967), among others, have recognized that one of the most critical pieces of evidence concerning the contribution of prenatal and postnatal influences to measured intelligence would be a comparison between twins brought up together and twins separated from birth. Evidence of this type was explored by McDonald (1964) in a small series from which results were somewhat inconclusive.

In the present investigation it has been assumed that twins were brought up together if both took the eleven-plus examination and that they were brought up like single births if their co-twins were stillborn or died within a month of birth. In this comparison standardization for maternal age and number of previous sibs has been necessary, since both variables have an influence on early mortality.

Twins brought up alone scored appreciably higher than those whose co-twins survived, although their birth weights were substantially lower (Table 9). The mean score of single surviving twins (98.8) was only 1.3 points below that of singletons (100.1). This difference is further reduced (to 0.7) when the mean score of singletons is adjusted by standardization (to 99.5) to the maternal age and birth rank distribution of twins. On this evidence, any increased difficulties associated with the prenatal development or birth of surviving twins have little or no influence on their verbal reasoning at age 11.

Table 9. Mean V.R. scores and mean birth weight of twins according to fate of co-twin

		Fate of co-twin		
		Stillborn or died in first 4 weeks (a)	Survived and sat examination (b)	Difference (a)–(b)
Males	V.R. score	98.2	94.4 (crude)	3.8
			93.9 (standardized*)	4.3
	Birth weight (kg.)	2.34	2.58	−0.24
		(85)	(976)	
Females	V.R. score	99.3	96.9 (crude)	2.4
			96.5 (standardized*)	2.8
	Birth weight (kg.)	2.22	2.45	−0.23
		(63)	(948)	
Total (giving equal weight to each sex)	V.R. score	98.8	95.6 (crude)	3.2
			95.2 (standardized*)	3.6
	Birth weight (kg.)	2.28	2.52	−0.24
		(148)	(1924)	

*Standardized to the maternal age/birth rank distribution of group (a). The mean score of single births standardized to this distribution was 99.5.

DISCUSSION

The intelligence of twins, as measured by V.R. scores in the eleven-plus examination, has a dual interest. It is of interest in its own right, as an indication of the degree of handicapping experienced by twins in consequence of the coexistence of two individuals before and after birth. But the twin data also throw light on some problems of single pregnancy, particularly interpretation of the variation in scores associated with birth weight and duration of gestation. This variation will be considered first.

The sib evidence previously examined (Record *et al.* 1969) showed little difference in V.R. scores in relation to birth weight; the twin data support this conclusion. Score differences between twins from the same pairs are trivial unless the weight differences are large (more than two-thirds of a kilogram) when there must be doubt about the comparability of the pair in respects other than weight.

But even the reservation concerning considerable weight differences seems to be overcome by the observations on twins whose co-twin died before or soon after birth. Such twins, raised singly, have a mean score which is only slightly lower than that of single births. This indicates that the large differences in weight and duration of gestation between single and multiple births do not result in appreciable score differences. It therefore seems justified to conclude that the smaller variations in weight and gestation length in single pregnancy also have little effect, and that the substantial score differences associated with these variables in a general population of births are a reflexion of other influences.

Since twins and their co-twins have a common pregnancy, their comparison can throw no light on the relation of intelligence to maternal age and birth order. But the data previously examined, particularly the sib evidence, suggested that the wide variation in association with these variables is largely a reflexion of other influences such as social class (Record *et al.* 1969*a*).

However, the present communication has been concerned mainly with investigation of the difference in verbal reasoning scores between twins and single births. The lower twin scores cannot be explained by differences from single births in maternal age and birth order or in birth weight and duration of gestation. They are not accounted for by the risks associated with monozygosity. These risks are reflected in lower birth weights, shorter gestations and higher stillbirth and neonatal death rates for like-sex than for unlike-sex twins (McKeown & Record, 1952; Record, Gibson & McKeown, 1952); yet V.R. scores are approximately the same for both. Scores are also unrelated to the order of delivery of twins in the same pair, although the second born is at greater risk than the first (Camilleri, 1963). Taken together these observations suggest that considerable variation in experience before and during birth has little influence on measured intelligence, and that the explanation of the substantial difference between twins and single births must be sought in the postnatal environment.

The same conclusion emerges from comparison of twins raised together with twins raised singly because of the death of the co-twin before or soon after birth. This comparison overcomes many difficulties of interpretation, by eliminating variation in prenatal experience and allowing attention to be focused exclusively on postnatal influences. This approach was thought to be sufficiently important to make it desirable to increase the number of twins from the period first investigated (1950–54) by adding those from three additional years (1954–57). The finding that twins raised singly have V.R. scores which are higher than those for twins raised together and almost equal to those of single births provides evidence that the handicapping of twins in respect of measured intelligence is determined after birth.

The verbal backwardness of twins has been a subject of interest in the psychological literature for many years. Although it has not hitherto been possible to exclude prenatal influences it has been suggested that the retardation may be due to the frequent contact between twin and co-twin, and reduced opportunities for verbal communication with adults and older sibs. This explanation, although still speculative, is consistent with the data presented here.

SUMMARY

Mean V.R. (verbal reasoning) scores recorded in the eleven-plus examination for Birmingham multiple births in the years 1950–57 were 95.7 for 2164 twins and 91.6 for 33 triplets. The mean for 48,913 single children born in the years 1950–54 was 100.1.

The low scores of twins are not explained by differences from single births in their distributions by maternal age and birth order or by birth weight and duration of gestation. They are also not accounted for by the increased risks associated with monozygosity (assessed by comparison of like- and unlike-sex twins) or with delivery of the second twin. Taken together these observations, like the previous ones on single births, suggest that variation in experience before and during birth has little influence on measured intelligence and that the explanation of the large difference between twins and single children must be sought in the postnatal environment.

There were 148 twins whose co-twins were stillborn or died within 4 weeks after birth; their mean score was 98.8, only a little lower than that of single births (99.5) standardized to the maternal age and birth rank distribution of twins. From this evidence it is concluded that the handicapping of twins, reflected in their low verbal reasoning scores, is due to postnatal rather than prenatal influences.

These conclusions are of course based on children who took the eleven-plus examination and cannot be accepted without reservations for those who did not.

Acknowledgements

We are greatly indebted to Dr E. L. M. Millar, Medical Officer of Health for Birmingham and Mr K. Brooksbank, Chief Education Officer for Birmingham, for permission to use the records of their departments. We should also like to express our thanks to Mr P. C. Fletcher of the Birmingham Education Department for his generous assistance in obtaining data for the past few years and to Mrs Karen Glenn for preparing computer programs.

References

Camilleri, A. P. (1963). In defence of the second twin. *J. Obst. Gynaec. Br. Cwth.* **70,** 258.

Churchill, J. A. (1965). The relationship between intelligence and birth weight in twins. *Neurology* (*Minneap.*) **15,** 341.

Day, E. (1932). The development of language in twins. Comparison of twins and single children. *Child. Dev.* **3,** 179.

Davis, E. A. (1937). The development of linguistic skills in twins, singletons with siblings and only children from age five to ten years. *Univ. Minn. Inst. Child Welf., Monogr.* No. 14.

Drillien, C. M. (1961). A longitudinal study of the growth and development of prematurely and maturely born children. Part VII: Mental development 2–5 years. *Arch. Dis. Childh.* **36,** 233.

Illsley, R. (1967). Family growth and its effect on the relationship between obstetric factors and child functioning. In *Social and Genetic Influences on Life and Death.* Ed. Lord Platt and A. S. Parkes. Edinburgh: Oliver and Boyd.

Lewis, M. M. (1963). *Language, Thought and Personality in Infancy and Childhood.* London: Harrap.

McCarthy, D. (1954). Language development in children. In *Manual of Child Psychology.* Ed. L. Carmichael. New York: Wiley.

McDonald, A. D. (1964). Intelligence in children of very low birth weight. *Br. J. Prev. Soc. Med.* **18,** 59.

McKeown, T. & Record, R. G. (1952). Observations on foetal growth in multiple pregnancy in man. *J. Endocrinol.* **8,** 386.

Mehrotra, S. N. & Maxwell, J. (1949). The intelligence of twins. A comparative study of eleven-year-old twins. *Popul. Stud.* **3,** 295.

Record, R. G., Gibson, J. R. & McKeown, T. (1952). Foetal and infant mortality in multiple pregnancy. *J. Obst. Gynaec. Br. Emp.* **59,** 471.

Record, R. G., McKeown, T. & Edwards, J. H. (1969*a*). The relation of measured intelligence to birth order and maternal age. *Ann. Hum. Genet. Lond.* **33,** 61.

Record, R. G., McKeown, T. & Edwards, J. H. (1969*b*). The relation of measured intelligence to birth weight and duration of gestation. *Ann. Hum. Genet. Lond.* **33,** 71.

Sandon, F. (1957). The relative numbers and abilities of some ten-year-old twins. *J. Roy. Stat. Soc.* **120,** 440.

Willerman, L. & Churchill, J. A. (1967). Intelligence and birth weight in identical twins. *Child Develop.* **38,** 623.

10. Genetics and Sociology: A Reconsideration

BRUCE K. ECKLAND

Ever since Mendel discovered that the cross-fertilization of a smooth-surfaced pea and a wrinkled pea had a predictable outcome, the ensuing issue of "nature versus nurture" in the development of human behavior has remained unresolved. Associated with the failure of the biological and social sciences to reach a resolution has been a general neglect of the *interaction* between genetic and social processes. Rather than considering our being as a product of nature *and* nurture, until recently the separate disciplines have seldom moved beyond their gainless attempts to attribute some specified amount of the observed variability to one source or the other.

Explanations of measured intelligence, for example, usually either followed the behaviorist's optimism, which denied any major role to heredity, or followed the equally extreme position of hereditarians who have consistently maintained that the influence of environment is quite small. In areas other than intelligence, the situation has been much the same. With few "moderate" spokesmen, the logical ties between genetics and the social sciences slowly deteriorated and nearly were buried.[1]

Moreover, as we shall note, sociologists have been far more resistant to any synthesis or working arrangement between the biological and social sciences than have other investigators in these areas, including the geneticists themselves. It appears, in fact, that the last major sociological work published in the United States that gave serious attention to the interdependence of heredity and environment was Sorokin's essay in 1927 on social mobility.[2]

We will argue for the advancement of a theory of the organization of social and genetic processes.[3] Not only under specified conditions can both environment and genetic variabilities be demonstrated, but geneticists, anthropologists, and psychologists have begun to articulate the connections between heredity and environment in new ways. Many of their concerns, moreover, do not deal strictly with genes, culture, or the individual, but with the traditional subject matter of sociology.

Reprinted by permission from *American Sociological Review,* 1967, Volume 32, No. 2, 173-194.

The discussion to follow centers on mental ability or "intelligence."[4] Although many of our conclusions may apply to other behavioral traits, our choice is a simple one. Some of the crucial issues in the evolution of modern societies concern the allocation of status, social mobility and the like; in this process the identification, development, and utilization of talent have become increasingly familiar elements. Probably few readers would disagree. Perhaps unclear, in light of this, is just how genetic and sociological principles may be brought together to produce a more adequate conceptualization and understanding of essentially sociological issues. The greater part of this paper will be devoted to this problem, with special reference to population genetics, the family, education, and social mobility.

Our objectives, then, are three-fold: (1) To review some of the conditions that have brought us to our present posture on the subject of genetics; (2) To suggest a number of reasons why we, as sociologists, should re-examine our posture; and (3) To describe four "sociological" problems in which the integration of genetic and social processes is especially relevant.

OUR PRESENT POSTURE

First, as our discipline has developed, we have had a vested interest in establishing a strong environmentalist approach to the study of human behavior. Moreover, we have been moderately successful in doing so. One of our favorite targets, for example, has been educational testing, especially IQ testing. There appear to be so many subtle ways in which environmental factors may contribute to inter-individual differences on these tests that some observers are nearly convinced that, except for a small number of mental defectives, no genetic component is involved whatsoever, at least in terms of what is being measured. Besides the early development of verbal skills, there is evidence that achievement motivation, formal academic training, and the immediate conditions of the test-taking situation itself all may differentially influence an individual's test performance.[5]

It should not be forgotten, too, that the sociologist's posture has been reinforced substantially in the past by many cultural anthropologists, some of whom still hold to the view that the biological evolution of man has run its course and cultural evolution has long since taken over. At one time, we not only gained the impression from them that man is no longer evolving in any genetic sense but that the genetic base of man is uniform everywhere. Carried one step farther, this often has been taken to mean that all observed variations in human behavior both within and between Mendelian populations are the sole result of cultural determinants.

Yet, have some cultural determinists and other ardent environmentalists carried their argument too far? Is the plasticity of man so unlimited? Is the genetic basis of man so uniform and, therefore, inconsequential? We will return to these questions in a moment.

A second condition that has led us to disregard the work of behavior geneticists seems to involve the utility of our own perspective. There have been few, if any, simple means for applying genetic principles in such a way as to produce a significant change in the cultural or social arrangements of human societies. Not only are technical problems involved but the very idea of controlling genetic processes brings forth the sinister and repugnant image of the "Brave New World." Perhaps for this reason even those who agree with the proposition that genetically "all men are *not* born equal" often overlook its significance.

In contrast, social arrangements can be altered by manipulating the environment with much greater convenience, with less resistance, and, to a large extent, without regard to genetic processes. This seems to be the position of many sociologists, like Faris, and others who have suggested that the present limits of our nation's supply of mental ability are not set by genetic factors but very appreciably by the environment.[6] We agree that there is considerable room for improvement in the training and utilization of our collective manpower. It also is easy to agree with Faris' conviction that "immense potentialities of human abilities are being smothered by systematic social influences." On the other hand, we cannot share his renewed optimism that "anybody can learn anything."[7]

Certainly, except for a small proportion of the population, it may be appropriate for some purposes to assume that all human beings are born with the potentiality to act intelligently. But what do we mean by acting intelligently? Does this include everything from the normal conduct of man's affairs to the kinds of decisions that might be involved in dropping the next atomic bomb? Is there no important genetic variability in human functioning beyond certain minimal (and, we believe, arbitrary) structural prerequisites, as some have suggested? This is a static concept of man and society. It also is a concept insensitive to the joint effects of the biological and social processes that undoubtedly are involved in the development of intelligence.[8]

Lastly, we would be amiss not to mention one other condition that has brought us to our present posture. This involves the traditional values which the social scientist, and the sociologist in particular, has held and still holds. On this and most other controversial issues he is a "liberal" or at least believes in a free, equalitarian society. However, fortunately or not, it is quite easy for this ideology, or any ideology, to influence the manner in which the scientist approaches his subject matter. A statement by a British psychometrician, P. E. Vernon, aptly describes what we mean:

> Those with left-wing opinions dislike the assumption that anyone born from an upper- or middle-class family has some innate superiority over those of less privileged birth, and believe that social reform and improved education will rectify such divergencies. Whereas the view often expressed in the nineteenth century, and still occasionally heard, is that the poor cannot benefit from, and do not deserve, as good an environment and education as the rich. Communist theory so strongly emphasizes the modifiability of genetic constitution that for a time Mendelian principles were rejected in Soviet Russia, and intelligence testing is still regarded merely as a ruse to perpetuate social-class and educational differences. [The most extreme hereditarians were the Nazi racial theorists of the 1930s, who ascribed all the desirable human traits and abilities to people of Nordic descent, all the undesirable ones to Jews.[9]]

There should be no forced choice here, since neither position, of course, is consistent with the evidence. On the other hand, even if neither of these positions could be rejected on presentation of a scientific argument, we, as sociologists, still would have no license to dismiss unpopular theories simply because they are inconsistent with popular ideologies.

WHY CHANGE?

> 'Inheritance of intelligence' does not mean that one's wits are decided and fixed at conception or at birth; it only means that with uniform upbringing and education people's wits would continue to be variable.[10]

Only in *quantity,* never in *fact,* has the contribution of heredity to the observed variability of human behavior ever been effectively challenged. Although environmental components are easily identified, they typically "explain" only a relatively small amount of the total variance in test (IQ) performance.[11] This does not prove the case for genetics any more than it disproves the case for environment, since the "unexplained" variance may involve a certain amount of measurement error, plus whatever environmental *and* genetic factors we may have neglected to consider.

First, in theory, geneticists do have a straightforward estimate of the mean proportion of genes that one has in common with any particular relative. Whether in reference to intelligence or other continuous traits, like height, the coefficients are exactly the same.[12] For example, between a parent and child or between ordinary siblings (including fraternal twins), roughly 50 percent of the genes are held in common.[13] Interestingly, the observed parent-child and sib-sib correlations on intelligence are quite orderly and generally obtain the theoretical value. When all genes are held in common, as in the case of identical twins, again the observed correlations in all studies have been generally in line with the theoretical value. The similarity in test performance of identical twins reared together, about 0.87, is nearly as high as the intra-individual reliabilities of the tests employed, i.e., the correlation between two parallel tests for the same individual.[14] While this contrasts with a lower correlation of about 0.75 in studies of identical twins reared apart, note how much higher even this latter figure is than the 0.23 correlation usually found for unrelated persons reared together.[15]

These comparisons illustrate what Dobzhansky meant when he observed that "whenever the matter has been studied, both genetic and environmental components of the variability have usually come to light."[16] As a result, behavioral geneticists have made a useful and important distinction between the concepts of genotype and phenotype. Whereas genotype refers to the presumably unalterable genetic make-up of an individual, phenotype refers to the observable physical and social characteristics that result from the interaction of the individual's genotype with his environment. Owing to environmental differences, it is often the case, as Gottesman notes, that "different genotypes may have the same phenotypes, and different phenotypes may be displayed by the same genotypes."[17] In other words, even though the theoretical genotypic similarity and observed phenotypic similarity are very close on mental tests, two individuals with the same IQ scores may have quite different genotypes, while another two individuals with quite different scores may, in fact, have very similar genotypes.[18] The phenotypic expression of intelligence, therefore, is always a relative and flexible matter; the genotype is not.

Although the related genotype(s) probably cannot be clearly identified by any phenotypic (behavioral) measure of intelligence, the genetic component nevertheless has been demonstrated rather convincingly in the twin studies. Most common have been comparisons between identical and fraternal twins.[19] In this situation it is presumed that because the trait-relevant environments are about the same for all twins who are reared together in the same family, the genetic component can be inferred from the greater similarity of performance between identical twins than between fraternal twins. An alternative, though less frequent, approach has been to compare identical twins who have been reared together with those who have been reared apart. Thus the same genotypes are studied in different environments, rather than different genotypes in the same environment.

In both types of studies, roughly 70 percent of the variance within families in intelligence has been attributed to genetic heredity. Moreover, the findings have been remarkably consistent despite differences in the methods used to construct the heritability coefficients, differences in the types of intelligence tests, differences in the age structure, ethnic composition, or socioeconomic character of the samples, differences in the regions studied, i.e., whether local, national, or foreign samples, and differences in sample size, which have been unusually large in some cases, such as the 631 twins in a recent Swedish investigation[20] and the 1169 twins in a study reported recently by the National Merit Scholarship Corporation.[21]

Given, however, the somewhat tenuous assumptions that all twin studies are required to make, the size of the heritability component perhaps has been exaggerated. Nevertheless, even allowing for some measurement error and questionable controls, the estimated variance attributed to heredity on these tests probably would remain very high. The factual evidence really is not in dispute.

In addition to the twin studies and the orderliness of other correlations between family members of varying degrees of relationship, the presence of a genetic component in conventional IQ tests has been demonstrated in studies of foster and adopted children. If intelligence is determined primarily by environment, it is argued, then the IQs of children removed from their biological parents as infants certainly should correlate more closely with the IQs of their foster parents than their biological parents. The inverse, however, is true.[22]

Again let us stress the point that inter-individual differences in test performance are not due solely to variations in heredity. We agree that most, if not all, intelligence tests consistently favor children reared in a stimulating and instructive environment. Yet, no research has demonstrated that the cultural component in these tests "explains" as much as 50 percent of the inter-individual variance; while, at the same time, no research has ever found that the genetic component "explains" less than 50 percent of the variance.

On the other hand, the evidence must be interpreted cautiously, not because it is wrong but because to ask how much behavior is determined by heredity and how much by environment is not a very sensible approach. Heredity and environment should not be set against each other in this way; as we shall note, the problem is far more complex. We merely wish to emphasize here that the acceptance of one set of evidence does not require us to reject the other. Rather, we should accept both, and once having done so, begin to move on to the more important work of developing a common framework for understanding the connections between the biological and cultural evolution of human societies.

There are also other reasons for re-examining our posture with regard to the nature-nurture question. One of these reasons stems from a marked change in the orientations of behavior geneticists and anthropologists to the problem. Geneticists, for example, have all but given up the question of how much a behavioral trait is due to heredity and how much to environment. Rather, many are now concerned with such questions as how modifiable by systematic manipulation of the environment is the phenotypic expression of a trait and how heredity interacts with the environment to produce phenotypic variations.[23] One approach to these questions is to think of heredity as determining a norm of reaction. Gottesman explains the approach in this manner:

Within this framework a genotype determines an indefinite but circumscribed assortment of phenotypes, each of which corresponds to one of the possible environments to which the genotype may be exposed. *If* and *when* the effort is expended to change radically the natural environment of a known genotype, e.g., maze bright and maze dull rat strains, there is a complete masking of the genotypic expectation and the two strains cannot be differentiated from their behavior . . . It is the plasticity of such a phenotypic character as human behavior in a heterogeneous environment and its invariance in a homogeneous environment which has led extremists to lose their perspective about the situations in which nature is more important than nurture and vice versa.[24]

Moreover, behavior geneticists, like sociologists, are interested in exploring the complexities of human behavior in such areas as child socialization, educational attainment, fertility, and social mobility. Their vantage point warrants our attention.

Many anthropologists, in dealing with evolutionary theories, also have changed their perspective. The "critical point" theory, which only ten years ago had stifled any real synthesis between physical and cultural anthropology, is being redressed in some quarters. The theory had held that the biological evolution of man proceeded up to a "critical point" when suddenly man became a culture-bearing animal. Having reached this "point," any further development of man's physical being became inconsequential as his accumulation of ingenious devices (culture) extended his pre-existing capacities like a "superorganic cake of custom."[25] The recent progress, however, of archeologists and paleontologists in their efforts to piece together the evolutionary record of man through human fossils and cultural artifacts has led to quite a different perspective. Rather than sequential, the biological and cultural evolution of man appears to have occurred in a slow synchronous (reciprocal) manner. In other words, the development of man's innate capacity was itself dependent upon the gradual development of culture.[26] This not only suggests that man's cultural environment probably will continue to shape his future physical development but that heredity and environment involve inseparable processes.

Finally, if for no other reason, we must begin to take note of genetic principles because, whether or not we or other social scientists accept them, their generally popular acceptance has very real consequences for social arrangements. Let us explain. Technological advancement, especially in a competitive world society, has stimulated us to utilize all human resources more effectively and to seek solutions to the problems that now obstruct their full development. Moreover, against an existing normative backdrop of equalitarian values, this force is translated into the ideology of a meritocracy.[27] While allowing for such contingencies as differences in individual performance, a complex division of labor, and differences in social rewards, the meritocracy is a society in which positions are allocated on the basis of "talent" (plus "effort") rather than class or social advantage. A rational utilization of talent, however, requires some degree of control over ongoing practices which, to be effective, in turn requires special knowledge about the distribution of talent and the mechanisms that enhance or impede its development. Thus, the "Great Society" has geared itself to achieve the fullest (and fastest) development of each individual *according to his capacities,* and, as the search for talent goes on, mass programs in mental testing become part of the establishment.

A basic and often explicit assumption is, of course, that the intellectual "endowments" of individuals *do* differ and that, owing to these differences, not everyone *can* learn anything. Furthermore, it is generally presumed, at least by those whose decisions are most likely to affect current practices, that attempting to act as if everyone *were* equal, as in uniform educational programs, is inefficient, impedes the development of the brighter student, and places unreasonable expectations on the probable achievement of the slow learner.

The significant questions then become: To what extent are mental tests actually used, for what purposes are they used, what do people believe the tests measure, and is their use generally accepted? Later we will elaborate on the answers to these questions. It suffices, for the moment, simply to note that students, teachers, and parents typically believe that the standardized tests now in use measure, at least to some extent, the traits with which a person is born. In addition, the general acceptance and use of these tests has become so widespread in recent years that it is inconceivable that their continued use will not have a very profound impact on the structure of the American society in the years ahead. Goslin, one of the principal investigators of this particular problem, has summarized the situation in the following manner:

> For the first time in the history of the world, a conscientious attempt is being made to measure objectively the intellectual abilities of human beings and to make it possible for those individuals having the greatest abilities to rise to positions of high status. Although many other characteristics of individuals still play an important role in the assignment of status in western society, objective estimates of ability are fast becoming of crucial importance as a result of the development of standardized ability tests.[28]

In the remainder of this paper, we will present four brief but specific illustrations of the manner in which genetic and social processes, if taken together, could provide the basis for framing an essentially sociological investigation. As we have need for them, genetic principles will be introduced, including those dealing with breeding populations, regression toward the mean, and random genetic drift.

CLASS, FERTILITY, AND INTELLIGENCE

In the classical treatment of social classes as Mendelian or "breeding" populations, attention has focused on conceivable alterations in man's genetic character due to a combination of differential fertility rates (favoring the lower class) and assortative mating, i.e., marriages between phenotypically, and therefore presumably genotypically, similar individuals.

A breeding population simply is an aggregate of individuals who are statistically distinct from other aggregates with respect to some gene frequencies as a result of assortative mating and other processes. Although the observed differences in IQ tests of children born of parents from different classes admittedly are attributable in part to cultural variations, the findings are at least equally predictable from a genetic theory of intelligence. It should be no surprise, therefore, that correlations between spouses are found to range between 0.3 and 0.6 in most studies, in terms of both measured intelligence and various socioeconomic characteristics.[29] Assortative mating then *does* occur.

Furthermore, although no direct evidence is available, assortative mating with respect to intelligence probably is increasing, owing to the growth of a mass system of higher education in which mental abilities are the main criterion for selection, as we will note, and where sex ratios are beginning to approach equality.[30] Social classes, most likely, have always been breeding populations; and today, to the extent that intelligence has a genetic base and social positions are allocated according to ability, assortative breeding between genetically similar adults tends to favor the reproduction of genetically similar children in any given stratum. As in Young's meritocracy, talented adults rise to the top of the social hierarchy and the dull fall or remain at the bottom. Therefore, as the system strives to achieve full equality of opportunity, the observed within-class variance among children tends to diminish while the between-class variance tends to increase on the selective traits associated with genetic differences.[31]

Yet the system never reaches full circle, or to the point where mobility ceases, because no parent ever reproduces an exact replica of himself. Human variability is just as much a law of nature as is human heritability. Without the former, the entire process of evolution (natural selection) is impossible; without the latter, no species, *qua species* could survive.

In viewing social classes as breeding populations, we hesitate to speculate upon the many sociological implications. Yet in the area of fertility and population studies, there are some rather firm and interesting data.[32] Differential fertility by socioeconomic status has been observed for centuries,[33] and until lately, the low fertility of high-status groups frequently has been referred to by sociologists as one factor that accounts for vertical mobility.[34] With the advent of intelligence testing, moreover, a similar inverse correlation has been found between family size and the IQ of children.[35] As a result, a number of writers several years ago predicted that the collective intelligence of the population would gradually decline at a rate of one to four IQ points each generation.[36] There was nothing especially complicated about their reasoning. If the least intelligent adults of the lower classes were producing proportionately more children, as they thought, and if living conditions had substantially improved life expectancies and, therefore, the reproductive capacities of the children of the lower-classes, then the total forces of selection were favoring lower intelligence.

Yet longitudinal data in Scotland,[37] England,[38] New Zealand,[39] and the United States,[40] while covering only about one generation, have shown no decline. The paradox finally was resolved when it was recognized that fertility and family size do not correlate with intelligence in the same manner.[41] Previous studies, it is noted, had surveyed children only and thereby excluded the unmarried and infertile adults. Accordingly, a number of studies recently have demonstrated that these adults have the lowest intelligence and, when taken into account, no simple linear correlation between fertility and intelligence exists.[42] In other words, although the least intelligent groups produce more children within marriage, they are the least likely to marry.

While these findings, along with a number of equally cogent arguments,[43] may temporarily put aside the ominous thesis that selective forces have been operating in an unfavorable direction for the species, they set forth another interesting question. If social class and intelligence are positively related and fertility and intelligence are *un*related how do we explain the negative correlation between fertility and social class? What might be suggested is either that the high-ability, high-status adults have more children than the high-ability, low-status adults or that the low-ability, high-status adults have more children than the low-ability, low-status adults.[44]

Either hypothesis appears defensible. For example, young intelligent adults of child-bearing age and from lower or lower-middle class backgrounds may anticipate upward mobility or at least recognize the utility of family planning and therefore limit their family size. On the other hand, family size probably does not especially restrict the opportunities of adults from higher socioeconomic groups.[45] In regard to the alternative hypothesis, dull, and especially retarded adults from lower socioeconomic families tend to be less fertile in part because, as we already have noted, they tend not to marry. In contrast, it is perhaps by no means unusual for an upper or upper-middle class family to encourage the marriage and fertility of a moderately dull offspring, while taking, in the process, whatever precautions necessary (especially in the case of a son) to ensure the "selection" of an "intelligent" mate. These hypotheses suggest some new directions for research.

THE FAMILY AND THE "REGRESSION TOWARD THE MEAN"

This illustration deals with the nuclear family in urban industrial societies. As an extension of the genetic and sociological principles already discussed, our hypothesis is that a basic contradiction exists between the institutionalization of child-rearing practices in the nuclear family and the parent-child nexus. By fortuitous circumstances, the family does not and cannot effectively perform its most primary societal function. The point certainly could be argued using the dialectics of a model of class, race, or sex discrimination. For purposes here, however, we will argue essentially from the standpoint of genetics, though no geneticist (or sociologist) would necessarily agree with our hypotheses.

The irreconcilable predicament of the modern family, from this standpoint, is the genetic variability of man about which we previously spoke. This variability has two important consequences here. It produces a standard regression toward the mean; and on the other hand, in large breeding populations, it produces a broad range of genotypes that even within a uniform environment could conceivably produce almost any known phenotypic trait.

In regard to the first point, the standard regression toward the mean undoubtedly is, in part, a statistical artifact that should be expected to occur when there is any chance for measurement error. That is, whether the same individuals are being re-tested or whether two groups of individuals are being compared, such as the IQs of parents and children, the scores that are being matched against persons who initially were placed high or low on the scale will tend to be closer to the mean. More important to our argument, however, is that regression toward the mean also occurs because this is the manner in which probabilities in genetics actually work.[46] In other words, although a parent who is above average in intelligence is likely to have children also above average, the "native" intelligence of these children, on the whole, will be lower than that of the parent—that is to say, toward the mean of the total adult population. The inverse, of course, is also true. The chances are greater for a dull parent to produce a brighter than a duller child; but usually the child will still be somewhat below average.[47]

In regard to the second point, the range of possible hereditary traits that are transmitted in any particular mating is limited only statistically. The overlap or variability in measured intelligence between breeding populations has always been a remarkable discovery, and certainly not easily explained from other than a genetic standpoint. How else *can* one explain the perhaps more than occasional exceptions of a bright child born of dull lower-class parents or a subnormal child born of bright middle-class parents? Although, again, some of the exceptions probably can be explained by "environment," we seriously doubt that all or even most exceptions can be explained in this manner. Even under the most restricted environment, it is possible for dull parents to produce a bright child, and vice versa. Moreover, it is entirely plausible that there are far more exceptions than actually observed, since culturally-induced responses probably mask *both* the brightness of some children and the subnormality of others. Interestingly, while most observers are quick to concur that a restricted environment may disguise the native intelligence of some lower-class children, we seldom inspect the reverse situation. That is, it is equally probable that an enriched environment, especially one which accentuates "social" skills, may tend to disguise the subnormal intelligence of some middle- or upper-class children.[48]

How then does genetic variability put some modern families in a predicament? The task of rearing a particular child is the "culturally-induced" responsibility of the parents who are directly related to the child but whose genetic endowments may be too poorly matched (to the child) to succeed.

If our national ideals include equality of opportunity and a full utilization of all human resources and if to pursue these goals the structure of the family is so designed that children eventually must be emancipated from their families of orientation, then the obligation of parents to rear a child on account of the accident of birth is an obviously obsolete (or at least contradictory) feature of a modern society.[49] Differential advantages and handicaps are rooted in the nature of the parent-child relationship. Obviously, we are speaking here of cultural deprivation. However, rather than the common discriminatory class barriers we usually think of, the source of deprivation is tied to the variable mental capacities of parents to provide an appropriate environment for a particular child to develop according to *his* capacities.

Without discussing the full ramifications of this problem, we only raise the question: How is it possible for two dull parents to stimulate a potentially bright child, to be aware of his personal problems, and to effectively guide his development? It is not possible, and, apparently, it is partly for this reason that through crash poverty and educational programs we have intruded upon the social structure of the lower-class family. Because we believe it is both an "equitable" and "efficient" utilization of human resources, we physically and emotionally remove the child from his parents, in some instances, in order to give him opportunities more commensurate with his abilities.

One might wonder, however, just when we will begin to apply the same logic to the "equitable" and "efficient" treatment of the subnormal child of normal, middle-class parents. Do well-meaning intelligent parents, for example, have the "right" to attempt to push their children to a level of performance the achievement of which would place them under excessive emotional strain? Are children, instead, sometimes pushed into deliquent or other marginal roles?[50] How apropos is Conant's first point in the summarization of his observations of the American high school:

> The main problem in wealthy suburban schools is to guide the parent whose college ambitions outrun his child's abilities toward a realistic picture of the kind of college his child is suited for.[51]

EDUCATIONAL TESTING AND GENOTYPIC VARIABILITY

The variability in intelligence is a phenomenon closely linked to but usually evaded in current discussions on transformations in mass education in this country. Our discussion here will not be directly concerned with the (intergenerational) source of this variability, which was at the root of the preceding problem. Rather, we will be concerned with a basic dilemma that arises as attempts are made to institutionalize a system of educational selection based upon the variability itself when, in fact, we cannot perfectly identify the genotypic variation of intelligence but only its phenotypic expression (what the tests measure). Before elaborating upon this issue, let us look at one of the important elements involved, i.e., educational testing.

The development of standardized tests of ability has been closely associated with and partly a response to the growth of secondary education prior to World War II and, more recently, the growth of higher education. Beginning in the primary grades and continuing through college, standardized tests are now being employed as the primary basis for sorting and selection among a very diverse student population for placement in a very diverse range of programs. The extent of this development is itself quite remarkable. Below, we have summarized some of the most relevant findings from recent studies by the Russell Sage Foundation[52] and Project Talent[53] on testing in the nation's elementary and secondary schools.

(a) Virtually all schools now use standardized tests. Moreover, a large majority of the secondary schools in this country plan to expand their testing programs in the near future.

(b) Standardized tests are used to assess the potential learning ability of students, in order to provide individualized instruction, and to guide students in their decisions about school curricula, going to college, and jobs. A fairly large majority of the nation's youth actually are being placed into homogeneous classes on the basis of these tests, either by establishing completely separate programs of instruction for different students ("tracking") or by assigning students to different sections of the same course ("grouping").

(c) A majority of both students and adults believe that standardized tests measure the intelligence a person is born with; although most, at the same time, recognize that learning makes an important difference. Students and parents alike believe that the tests are basically accurate. While teachers essentially agree with the students and parents on these points, they also tend to believe that the tests are their best single index of a student's intellectual ability.

(d) Although public acceptance of standardized testing seems to depend largely upon the specific purposes for which the tests are used, a majority of the students believe that these are and should be important, especially in terms of deciding who should go to college.

This is only part of the story. For the college-going aspirants, the first two major hurdles—Educational Testing Service's Preliminary Scholastic Aptitude Test (PSAT) and National Merit Scholarship Corporation's Qualifying Test—usually come in the eleventh grade. Scores on the PSAT are not used to decide *who* goes to college, since that decision presumably was made some years earlier. Rather, one of the primary purposes of the test is to help decide who goes *where* to college by giving each student an early estimate of the probabilities of being admitted to any particular school.[54] The National Merit examination, on the other hand, is designed to select among the uppermost two percent of the country's students those who will receive special commendation and awards.[55] While in competition with other programs, the growth of the PSAT and National Merit programs in the past ten years has been so phenomenal that both are being administered today in roughly three-fourths of the nation's high schools.[56]

The next major, and perhaps most important, challenge comes about a year later—the college admissions tests and, for those in accelerated programs, the Advanced Placement examinations. Whether administered nationally or locally, tests are being used today by most colleges in their admissions process or for placement. Indeed, the proportion of high-scoring students that a college can attract has become the most objective criterion available for ranking the colleges themselves.[57] In addition, many secondary schools and colleges are participating in the Advanced Placement Program of the College Entrance Examination Board. Although presently involving fewer students and schools, its growth has closely paralleled that of the National Merit program.[58]

As the student progresses further, the significance of educational testing does not diminish. Upon completion of most undergraduate programs, admission to nearly any reputable graduate or professional school is fast becoming dependent, in part, upon the student's sophistication on such nationally administered tests as the Graduate Record Examination, the Admission Test for Graduate Study in Business, or the Law School Admissions Test. Likewise, entry into a particular field direct from a four-year undergraduate program sometimes requires additional testing, such as the National Teachers Examination of the Foreign Service Examination. Even the Peace Corps requires the Peace Corps Entrance Tests.[59]

Now, the basic issue in educational testing and selection, we believe, involves two competing sets of forces. On the one hand, the pressures *to* test appear irreversible.[60] It is generally recognized by most observers that individual capacities to learn vary rather markedly, especially when it comes to complex subject matter for any given grade level.

Because it is both "equitable" (see conclusions) and "efficient" (as noted earlier), we test. We do so, too, and perhaps more importantly, because enrollment pressures have forced our colleges to become more selective, especially since the post-war babies "came of age."[61] This, in turn, has pushed testing and selection down the grades to a point where now a student, or really his parents and teachers, probably must decide sometime before he leaves the eighth grade whether or not he is going on to college.

We therefore institutionalize programs for the separate and individualized treatment of the slow learners, the average, and the "gifted." Homogeneous grouping or tracking within a school, differentiated programs between the various units of a school system,[62] or even the differentiation between the "public" and "private" sectors[63] of American education, all add up to the same thing. The use of standardized tests and differential treatment can be found at nearly every point in the educational system. Moreover, there is constant pressure, as noted, to introduce mental tests into the selection process at an increasingly early point in the child's life in order that the development of his capacities will not be left to circumstance.[64]

On the other hand, there are obvious dangers in the use of standardized tests which lead their users to proceed cautiously. These dangers, in part, directly involve the size and nature of the discrepancy between innate intelligence (the genotypes) and what the tests measure (phenotypes). In early childhood and infancy, when intelligence tests are least blurred by culturally-induced responses, they are least reliable (temporally unstable) owing presumably to the slow development of the neurological structures related to intelligence.[65] Yet in later childhood and adolescence, when the tests are more reliable (stable), they are less valid in the sense that the phenotypic expression of intelligence at later age levels is far more sensitive to culturally-induced responses.

The "resolution" of the problem (no real solution is presently possible nor teleological explanation intended here) has been to seek a balance between testing programs that are introduced neither too early nor too late. If premature, the determination of status rests on arbitrary (unreliable) data; if postponed, the advantages of individualized treatment are neglected, and, moreover, depressing environmental influences become firmly established and are all too conspicuous.[66]

Some educational programs appear to compensate for the "mistakes" that occur. One of the objectives of the open door junior college, for example, is to give the slow learners and culturally deprived a realistic opportunity to recoup their losses and, if it should exist, demonstrate their capability to do college work.[67] However, perhaps with the exception of innovations like Head Start, most of the established programs unfortunately attack the problem at a point when few of the even capable students can be retrieved from the involuntary course upon which they have been set. Thus, despite all efforts to keep the doors open, to keep the dropouts in, and to keep the testers out, "sponsored mobility" is more a part of the American scene than most observers probably have thought.[68]

We believe that more sociologists should put aside their aversion to mental testing and strive to understand and disentangle the problems set forth here.[69] If not, we may soon find, if it is not already true, that about half of the children by age seven (the current figure in England and Wales)[70] have already been "tracked" on ability lines. We do not mean to suggest that the "professional" opinions of social scientists or even politicians would matter very much, at the moment, since certainly the feelings of the educators and school teachers are far more likely to determine just what goes on in the classroom anyway. But perhaps this is because, in contrast to the teachers who claim to "know better," we continue to insist that "all kids are alike," except, of course, our own or except those who get to college where *we* can observe them and readily "explain" the observed variations on the basis of the diverse educational experiences they already have encountered.[71]

THE CONCEPT OF "PERFECT MOBILITY"

Our final illustration of the importance of genetic processes to the sociologist deals with the manner in which we typically conceptualize vertical mobility. Nearly all studies in this area have taken person-to-person comparisons of the social stratum occupied by fathers and sons as the unit of analysis, and then assessed the openness of the class structure either directly from these comparisons or by various kinds of indices of association (and dissociation).[72] Essentially, once having accounted for "forced" mobility, i.e., mobility which is caused by changes in stratum composition or size, the remaining movement or degree of association is taken as an estimate of "pure" mobility, or "equality of opportunity." For instance, in the standard mobility matrix, "inequality" is signified when the occupational distributions of the sons born of fathers in different strata do not match. Conversely, in the ideal model, a system of full equality is perceived as one in which identical proportions of sons from all strata eventually will enter any given occupation.

What is basically wrong with this model? Although individual differences in ability are acknowledged, at least implicitly, and although the model permits migration from one stratum to another on the basis of these differences, it assumes that ability is randomly distributed at birth and that any differences observed among the children of different strata are solely a matter of environmental conditioning. Any genetic basis for these differences is dismissed as irrelevant. To dismiss the genetic factor, however, we would have to accept the unrealistic idea that, biologically, there is no more resemblance between a child and his parent than between a child and a total stranger.

If, on the other hand, an important part of the observed variations in intelligence between children from different social classes actually can be attributed to assortative mating and genetic processes, our model of "perfect" mobility requires a fundamental revision. That is, as mental functioning is made the principal criterion for the ascription of status, a model that specifies that full equality is achieved only when the same proportions of children from each class are assigned to any given status is not exactly appropriate. The matter-of-fact reason is that it apparently is becoming increasingly unlikely that the same proportions of children from each class have equal capacities to take advantage of their opportunities. The tendency for elites to replace themselves (intergenerationally) is somewhat ensured by the nature of any system in which intelligence is a dynamic factor affecting status placement.[73] Note that this particular restriction on mobility is conceptually distinct from all cultural barriers that may otherwise limit the channels of mobility and with which "equal opportunity" models are more properly concerned.

Lest the preceding discussion should be misunderstood, let us make firm the equally important point that genetic variability guarantees a relatively substantial pool of very bright children from lower-class backgrounds, which is sufficient reason, in both utilitarian and ideological terms, for social policies that emphasize the importance of this "overlap." Moreover, when entire groups of children from these backgrounds are systematically deprived of the same opportunities for development as other children, as presently persists, then the size of this pool is larger than that which otherwise might be estimated on the basis of heritability coefficients alone.

For instance, consider a hypothetical model of a simple bipartite structure in a moderately fluid system which has a one-to-three ratio between the size of the upper and lower classes. Although the upper class might produce the largest relative proportion of children with high intelligence, the greatest absolute number probably would come from the lower class.[74] This does not, however, dislodge our previous point; that is to say, the distribution of intelligence is by no means random and any mathematical or other model that purports to estimate the "openness" of a class structure should make adequate allowance for genetically-based inequalities.[75] These inequalities are not the same as the moral or political inequalities with which contemporary theories of stratification have generally been concerned.[76] Moreover, we question whether they should be, since equality of opportunity neither presupposes nor promotes equality of ability.

SUMMARY AND CONCLUSION

It appears that sociologists have tended to pass over some perfectly reasonable ties between their discipline and behavior genetics. They have done so, in part, because sociology, as a natural point of order, has had first to establish and articulate an environmental theory of behavior. At this, it not only has been successful, but there is reason to believe that its ideas can be and have been put into practice. Furthermore, sociological theories, intentionally or not, have been consistent with liberal-democratic values whereas genetic theories, in the past, have aroused far more suspicion.

We have argued, however, that the dichotomy between genetically and environmentally determined intelligence can no longer be permitted to dominate our thinking, since both really are parts of an interacting system. Our review of some of the recent evidence should begin to convince us that both heredity and environment are important and that neither agent alone can produce intelligence. Moreover, we noted that scientists in allied fields have a more balanced view of the question than do sociologists. Others appear willing to accord ample weight to environmental influences, but not to the exclusion of the genetic materials which, in important ways, regulate the adaptation of the human organism to its social environment, and vice versa.

To illustrate these connections, we presented four specific ways in which genetic processes are sociologically relevant. Although there are no simple solutions to the problems we outlined, it was suggested that we should be more sensitive to the work of behavior geneticists and avoid systems models, such as those in our current treatment of vertical mobility, that assume a uniform or random genetic base. We cannot opt for a biologically based model of social structure; on the other hand, we should not be so shortsighted as to continue to insist that "all men are created equal." While "inequality" is essentially a social phenomenon, it is nevertheless dependent upon the necessities of both social and genetic differentiation.

In concluding, we also can offer no special methodological tools for estimating the "norm of reaction" of the genotypes associated with intelligence. There are, of course, no means presently available for doing so.[77] Yet we need not wait for geneticists or others to provide more accurate instruments. Our present testing devices, including IQ, aptitude, or even achievement tests could be used, if cautiously, quite profitably in sociological investigations, and certainly far more than they are presently. Even though they do not isolate innate ability, individual and (sometimes) group differences on these tests do, in fact, measure it. Regardless, performance on these tests is swiftly becoming one of the most, if not *the* most important single criterion for the allocation of status.

All sociological investigations and commentaries, of course, have not neglected the treatment of intelligence. In addition to Anderson, Clark, Faris, and Rogoff, whose activities we already have noted, there are other sociologists who have dealt with the problem in a serious way. Of more than cursory interest are:

(a) Cicourel and Kitsuse's study of the professionalization of the role of the guidance counselors who administer the school's testing program and, owing to moral and bureaucratic imperatives, must "do something" when the academic performance of students falls below their predicted grade-point averages;[78]

(b) Goslin's continuing work on the social and psychological consequences of educational testing, and especially his interpretation of causal patterns of this form, the dissemination of test information→self-concepts of ability→achievement motivation;[79]

(c) Sewell's several studies on the interrelationships between neighborhood context, intelligence, and social mobility, one of which very recently engaged him (and Armer) in an important controversy in this journal when Turner, Boyle, and Michael all refused to accept the causal priority of intelligence when estimating the effects of neighborhood on educational aspirations;[80]

(d) Duncan's current working papers on how intelligence might be represented in a formal model of the process of achievement, and, more precisely, on how intelligence and "environment" might interact in this process when two separate estimates of the correlates of "early" and "later" intelligence are available;[81] and

(e) Farber's re-analysis of Burt's data using a procedure based on the Pascal triangle for identifying different effects of social class on intellectual development at the upper and lower IQ ranges, which indicates that the relationship between social class and intelligence is nonlinear, and suggesting that we must proceed cautiously when attempting to explain the interaction between heredity and environment.[82]

Also, in concluding, there is one point with which perhaps we should have begun, since it may help clarify the firm resistance of some sociologists who hold that any biological interpretation of human behavior is simply specious reasoning. We have in mind the "voluntarism" found in Linton's classic distinction between achieved and ascribed status[83] and its extension in the sometimes "soft" determinism of modern sociology.[84] In the meritocracy, as within any truly "deterministic" framework, Linton's distinction tends to break down, at least in terms of what we believe its most common usage to be.

Contrasted with "ascription," "achievement" generally refers to the allocation of status on the basis of properties not assigned by birth, such as class, sex, or race, but ones over which the individual presumably has some control and, therefore, "merits." In other words, the individual's *private* capacities are involved. But do not the individual's capacities depend to more than a trivial degree upon the genetic material with which he enters the social contest, and over which he has no more control than his race or his sex? And, therefore, is not the allocation of status according to ability actually just as much an "ascribed" criterion as the more traditional assignment of positions based on "social" heredity?

If one of the major social issues facing contemporary societies, as we have suggested, involves a basic confrontation between the principles of social heredity and the meritocracy, then which mode of selection is more equitable? Which is more "just" if volition is involved in neither the mental capacities that an individual inherits nor the social advantages conferred upon him by his parents?

Although one mode is perhaps no more equitable than the other, *one* does appear to be more rational. Here, we would agree with Linton's position, and others, that social heredity, while not "dysfunctional" in simpler societies, no longer meets the demands of a complex technology. To the extent that the survival of our present technology (and social order) depends upon the effective utilization of human resources, then the identification, sorting, and development of talent will continue to be persuasive arguments.

Notes

1. This paper is directed particularly to those who believe that the ties between sociology and genetics either "have been" or "should be" buried. For others, it is an attempt to illustrate the manner in which genetic principles might find their way into the sociologist's repertoire.

2. Pitirim A. Sorokin, *Social and Cultural Mobility,* New York: Free Press of Glencoe, 1964.

3. Fuller once advocated a similar position with respect to the subject matter of psychology, but he failed to illustrate how any synthesis between the hereditarians' and environmentalists' positions might be brought about. See John L. Fuller, *Nature and Nurture: A Modern Synthesis,* New York: Doubleday, 1954.

4. The term "intelligence" has no precise meaning. Although most psychometricians probably do not subscribe to Spearman's single-factor theory, only a few have begun to identify any of the conceptually distinct abilities that may be involved. The only distinction employed in this paper is between "innate" intelligence and that which intelligence tests generally measure, or, to be explained later, the genotypes and phenotypes of intelligence. The meaning hopefully in all instances, will be clear by the context in which the term is used.

5. Goslin, for example, has developed an interesting paradigm for studying the interrelationships of the influencing factors. See, David Goslin, *The Search for Ability,* New York: Russell Sage Foundation, 1963, pp. 130-132. A classical review of the original research on intelligence as related to socio-economic differences may be found in Kenneth Eells, Allison Davis, *et al., Intelligence*

and Cultural Differences: A Study of Cultural Learning and Problem-Solving, Chicago: University of Chicago Press, 1951. These and other correlates of intelligence, such as basal metabolism rates, EEG alpha frequency, etc., are reviewed by both Anastasi and Tyler; see Anne Anastasi, *Differential Psychology,* New York: Macmillan, 1958; and Leona E. Tyler, *The Psychology of Human Differences,* New York: Appleton-Century-Crofts, 1965.

6. In his presidential address in 1961 before the American Sociological Association, Faris gave this interesting analogy:

   "If it appears illogical to claim that physiological differences exist, but do not produce differences in performance, consider the rates of speed of automobiles on crowded metropolitan streets. The vehicles differ in horsepower, and in observed speeds, but the speeds may depend entirely on factors other than the horsepower—openness of the way ahead, urgency of the trip, nerves of the driver, and disposition of back-seat passengers."

   See Robert E. L. Faris, "The Ability Dimension in Human Society," *American Sociological Review,* 26 (December, 1961), p. 838, fn. 7. Faris later admits, however, that if the way ahead were open, if the trip were urgent, and if the driver and backseat passengers were similarly disposed, then the variability of the automobiles' horsepower has to be taken into account. Now in respect to the relationship between *genetic* horsepower and *human* performance, it does, in fact, appear that the structure of opportunities or "the way ahead" is becoming increasingly open, that achieving maximum performance is a goal which is being pursued with considerable urgency, and that the dispositions of everyone in the contest, although occasionally irregular, do not differ markedly in the sense that most want to "get ahead." If generally true, then it would also appear that the observed variability in performance has become increasingly dependent upon individual differences in the mental capacities that unavoidably handicap the slow learners and, just as unavoidably, favor the really fast ones.

7. *Ibid.,* p. 838. This optimism is as groundless as that of J. B. Watson, in the 1920's the leader of the school of behaviorism in psychology. K. B. Clark more recently has argued, too, that IQ scores indicate almost nothing about the ceiling of what an individual can learn but rather the rate of learning and the amount of effort that

learning will require; see Kenneth B. Clark, "Educational Stimulation of Racially Disadvantaged Children," in A. Harry Passow, ed., *Education in Depressed Areas,* New York: Columbia University, 1963. For the bulk of the formal subject matter in most elementary and secondary schools, Clark perhaps is right. His remarks, however, were made primarily in reference to group differences in IQ by race, and it is uncertain that he would carry this argument to the point of suggesting that with unlimited time and effort almost anyone can learn anything. Even if this were true, the rate of learning itself is a variable probably just as significant to the issues set forth in this paper as any inter-individual differences in the limits of learning. Just as the value of a computer depends largely upon the speed at which it can function, so too it would seem that human performance is closely associated with the rate of learning. In strictly economic terms, moreover, since the productive years of the life of any individual are biologically limited, there is far more return on the educational investment if it takes (only) 30 years, rather than 35, 40, or more, to train a scientist to perform at a minimum level of proficiency. Besides, limiting performance to any particular set of standards is hardly consistent with the rising and continuously changing demands of a modern technology.

8. While insisting that intelligence is neither fixed nor predetermined by heredity (with which we agree), after Hunt's very comprehensive review of the evidence he nevertheless insists that the genes prescribe basic directions in intellectual growth and set irrevocable limits on the range of capacities that can be developed (with which we also agree); see J. McVicker Hunt, *Intelligence and Experience,* New York: Ronald Press, 1961.

9. P. E. Vernon, *Intelligence and Attainment Tests,* London: University of London Press, 1960, p. 138. Cf., Sorokin, *op. cit.,* p. 330, nearly thirty-five years earlier; and John W. Gardner, *Excellence,* New York: Harper and Row, 1961, p. 56.

10. Theodosius Dobzhansky, *Mankind Evolving,* New Haven: Yale University Press, 1962, p. 86.

11. Occasionally, quite striking improvements or rises on IQ tests have been noted when retarded children or others whose background has been extremely adverse are placed in an "enriched" environ-

ment. Usually, however, the results are rather small, especially for any substantial batch of cases. Moreover, changes in test results under controlled conditions do not substantially reduce the variability of test performance, but only tend to move the entire range of scores upward. Obviously, these changes demonstrate that the environment is an important determinant; they do not, however, demonstrate that it is the only or necessarily even the most important determinant. Even Bloom, who in many ways is a leading proponent of the environmentalist view, after a careful review of the relevant materials, regarded (only) 20 IQ points as a fair estimate of any long-term effects that extreme environments might have on intellectual growth (although occasionally some of the "observed" changes exceed this figure). See Benjamin S. Bloom, *Stability and Change in Human Characteristics,* New York: Wiley, 1964, p. 71.

12. David Krech and Richard S. Crutchfield, *Elements of Psychology,* New York: Knopf, 1958, p. 576.

13. The theoretical additive parent-offspring and sib-sib correlations actually are 0.5 plus the quotient of the assortative mating coefficient divided by 2, and not 0.5 as commonly reported in the literature. See R. Stanton, "Filial and Fraternal Correlations in Successive Generations," *Annals of Eugenics,* 13 (1946), pp. 18-24. I wish to thank zoologist Carl Jay Bajema for drawing my attention to this point.

14. Although preschool-age IQ tests are notably unreliable, later correlations normally are quite stable over time. One twenty-five year longitudinal study of 111 individuals obtained the following correlations between test and re-test: preschool and adolescence (a ten-year span), 0.65; preschool and adulthood (a twenty-five year span), 0.59; and adolescence and adulthood (a fifteen-year span), 0.85, which again is nearly as high as the reliability of these tests. See Katherine P. Bradway and Clare W. Thompson, "Intelligence at Adulthood: A Twenty-Five Year Follow-Up," *Journal of Educational Psychology,* 53 (February, 1962), pp. 1-14.

15. The three statistics reported above are based on Erlenmeyer-Kimling and Jarvik's comprehensive review of this literature, covering over 50 studies and, in total, yielding over 30,000 correlational

pairings. See L. Erlenmeyer-Kimling and Lissy F. Jarvik, "Genetics and Intelligence: A Review," *Science,* 142 (December 13, 1963), pp. 1477-1479.

16. Dobzhansky, *op. cit.,* p. 322.

17. Irving I. Gottesman, "Bio-Genetic Perspectives," Duetch and Jenson, *Race, Class, and Psychological Development,* New York: Holt, Rinehart and Winston, in press. Environment is taken here not only to refer to the social milieu but to a host of prenatal or molecular factors between the embryonic cells. See C. H. Waddington, *The Strategy of Genes,* New York: Macmillan, 1957; and W. W. Meissner, "Functional and Adaptive Aspects of Cellular Regulatory Mechanisms," *Psychological Bulletin,* 64 (September, 1965), pp. 206-216.

18. Authors do not agree about the nature of the genotypes involved in intelligence. For example, Hayes provides a cogent argument that the hereditary basis of intelligence consists of motivational drives, rather than any specific or general abilities, as such. See Keith J. Hayes, "Genes, Drives, and Intellect," *Psychological Reports,* 10 (April, 1962), pp. 299-342. While I have not done so elsewhere, I will intentionally avoid, where possible, any specific discussion of motivation or values in this paper which, in contrast to Meissner's view, may be treated simply as part of the environment or social milieu. See Bruce K. Eckland, "Social Class and College Graduation: Some Misconceptions Corrected," *American Journal of Sociology,* 70 (July, 1964), pp. 36-50, and "College Dropouts Who Came Back," *Harvard Educational Review,* 34 (Summer, 1964), pp. 402-420.

19. Identical or "monozygotic" twins are produced from the division of a single fertilized egg, while fraternal or "dizygotic" twins arise from the fertilization of two separate ova. The latter are no more genetically alike than ordinary siblings. Thus, the chances are about even, for example, that fraternal twins will be of the same or opposite sex, whereas identical twins are always of the same sex. Most, although not all, identical twins can be diagnosed on the basis of appearance. In fact, recent cross-validation studies (against the more precise method of blood typing) have found that this approach is about 93 percent accurate. See Robert C. Nichols,

"The National Merit Twin Study," in Steven Vanderberg, ed., *Methods and Goals in Human Behavior Genetics,* New York: Academic Press, 1965. The point is noteworthy since one of the most frequent criticisms of the early twin studies was the questionable validity of similarity of appearance as a method for diagnosing zygosity.

20. T. Husen, *Psychological Twin Research,* vol. 1, Stockholm: Almqvist and Wiksell, 1959.

21. Nichols, *op. cit.*

22. Marie Skodak and Harold M. Skeels, "A Final Follow-Up Study of One Hundred Adopted Children," *Journal of Genetic Psychology,* 75 (September, 1949), pp. 85-125. Again, the evidence is subject to criticism, but mainly in terms of interpreting the relative strength of the genetic component, not its presence. For critical reviews, see R. S. Woodworth, *Heredity and Environment: A Critical Survey of Recently Published Material on Twins and Foster Children,* Social Science Research Council, Bulletin 47, 1941; and Anastasi, *op. cit.*

23. An introduction to the complexities of these questions can be found in the writings of the behavior geneticists, such as J. Fuller and W. Thompson, *Behavior Genetics,* New York: Wiley, 1960; Irving I. Gottesman, "Genetic Aspects of Intellectual Behavior," in Norman Ellis, ed., *Handbook of Mental Deficiency: Psychological Theory and Research,* New York: McGraw-Hill, 1963, pp. 253-296; J. Hirsch, "Individual Differences in Behavior and Their Genetic Basis," in E. Bliss, ed., *Roots of Behavior,* New York: Harper, 1962, pp. 3-23; and G. E. McClearn, "Genetics and Behavior Development," in M. L. and Lois W. Hoffman, eds., *Review of Child Development Research,* Vol. 1, New York: Russell Sage Foundation, 1964, pp. 433-480.

24. Gottesman, "Bio-Genetic . . . ," *op. cit.*

25. Clifford Gertz, "The Growth of Culture and the Evolution of Mind," in Jordon M. Scher, ed., *Theories of the Mind,* New York: Free Press of Glencoe, 1962, pp. 713-740.

26. For a recent and sensitive discussion by a zoologist on the convergence of organic and cultural evolution in the foundation of human

societies, see Alfred E. Emerson, "Human Cultural Evolution and Its Relation to Organic Evolution of Insect Societies," University of Chicago, unpublished manuscript.

27. Michael Young, *The Rise of Meritocracy,* London: Thames and Hudson, 1958.

28. Goslin, *The Search . . . , op. cit.,* p. 189.

29. For an extensive review of the literature on assortative mating, see J. N. Spuhler, "Empirical Studies on Quantitative Human Genetics," in *The Use of Vital and Health Statistics for Genetics and Radiation Studies,* United Nations and World Health Organization, 1962, pp. 241-252. Although Burt's interpretation of his own data is subject to criticism, he is, no doubt, one of the leading proponents of the idea that class differences in intelligence are largely due to genetic variation. See Cyril Burt, "Class Differences in General Intelligence: III," *British Journal of Statistical Psychology,* 12 (May, 1959), pp. 15-33, and Cyril Burt, "Intelligence and Social Mobility," *British Journal of Statistical Psychology,* 14 (May, 1961), pp. 3-24. Also see Bernard Berelson and Gary A. Steiner, *Human Behavior: An Inventory of Scientific Findings,* New York: Harcourt, Brace and World, 1964, p. 309. For general reviews on the relationship between social class and intelligence, see John B. Miner, *Intelligence in the United States,* New York: Springer, 1957; B. G. Stacey, "Some Psychological Aspects of Inter-Generation Occupational Mobility," *British Journal of Social and Clinical Psychology,* 4 (December, 1965) pp. 275-286; and Fuller, *Nature and . . . , op. cit.*

30. I base this proposition largely on the fact, too, that the education of two spouses accounts for far more of the variance in mate selection (and has for at least 30 or 40 years) than any other known factor. See, Bruce L. Warren, "Multiple Variable Approach to the Assortative Mating Phenomenon," *Eugenics Quarterly* (in press).

31. Dobzhansky, *op. cit.,* p. 244.

32. While dealing with these materials only in an introductory fashion here, more complete coverage can be found in Anne Anastasi, "Intelligence and Family Size," *Psychological Bulletin,* 53 (May, 1956), pp. 187-209; Anne Anastasi, "Differentiating Effect of Intel-

ligence and Social Status," *Eugenics Quarterly,* 6 (June, 1959), pp. 84-91; Cyril Burt, *Intelligence and Fertility: The Effect of the Differential Birth Rate on Inborn Mental Characteristics,* London: The Eugenics Society and Cassell, 1952; and J. N. Spuhler, "The Scope for Natural Selection in Man," in W. J. Schull, ed., *Genetic Selection in Man,* Ann Arbor: University of Michigan, 1963, pp. 1-111.

33. See Frank W. Notestein, "Class Differences in Fertility," *Annals of the American Academy of Political and Social Science,* (November, 1936), pp. 1-11, for a review of this subject.

34. The data no longer suggest today that the upper classes are not reproducing themselves.

35. The correlations cluster around –0.3; see Anastasi, "Intelligence and . . . ," *op. cit.* It should be noted, too, that the correlations persist when occupational class is controlled.

36. For references, see Gottesman, "Bio-Genetics . . . ," *op. cit.* and Vernon, *op. cit.*

37. Scottish Council for Research in Education, *The Trend of Scottish Intelligence,* London: University of London Press, 1949.

38. Raymond B. Cattell, "The Fate of National Intelligence: Test of a Thirteen-Year Prediction," *Eugenics Review,* 42 (October, 1950), pp. 136-148; and W. G. Emmett, "The Trend of Intelligence in Certain Districts of England," *Population Studies,* 3 (March, 1950), pp. 324-337.

39. Betty M. Giles-Bernardelli, "The Decline of Intelligence in New Zealand," *Population Studies,* 4 (September, 1950), pp. 200-208.

40. R. D. Tuddenham, "Soldier Intelligence in World Wars I and II," *American Psychologist,* 3 (1948), pp. 54-56.

41. With limited data at hand, Penrose appears to be the first to have recognized the possibility that persons of extremely low intelligence, whom he found more likely to be infertile, could balance out the decline attributable to their larger numbers. See L. S. Penrose, "Genetical Influences on the Intelligence Level of the Population," *The British Journal of Psychology, General Section,* 15 (March, 1950), pp. 128-136.

42. The results actually indicate a slight bimodal or curvilinear relationship, wherein low fertility is associated with both low and high (but not average) intelligence. See J. Higgins, Elizabeth W. Reed, and S. C. Reed, "Intelligence and Family Size: A Paradox Resolved," *Eugenics Quarterly,* 9 (June, 1962), pp. 84-90; and Carl Jay Bajema, "Estimation of the Direction and Intensity of Natural Selection in Relation to Human Intelligence by Means of the Intrinsic Rate of Natural Increase," *Eugenics Quarterly,* 10 (December, 1963), pp. 175-187.

43. Otis Dudley Duncan, "Is the Intelligence of the General Population Declining?" *American Sociological Review,* 17 (August, 1952), pp. 401-407; and Hunt, *op. cit.,* pp. 337-343.

44. Or perhaps, much more simply, the solution for this problem is quite the same as for the original paradox. That is, just as the low fertility of adults of very low intelligence was obscured in earlier studies, the low fertility of very low socio-economic (including institutionalized) adults may have been obscured in studies purporting to demonstrate the relationship between fertility and social class.

45. There is evidence, for example, of a negative correlation between upward intergenerational mobility and family size. See Dudley Kirk, "The Fertility of a Gifted Group: A Study of the Number of Children of Men in *Who's Who,*" in Proceedings of the 1956 Annual Conference of the Milbank Memorial Fund, *The Nature and Transmission of the Genetic and Cultural Characteristics of Human Populations,* Milbank Memorial Fund, 1957, pp. 78-98; and especially Richard F. Tomasson, "Social Mobility and Family Size in Two High-Status Populations," *Eugenics Quarterly,* 13 (June, 1966), pp. 113-121. Also, there is evidence that the negative association of fertility with general socioeconomic status masks its positive association with income, so that when other background characteristics like education (and we might venture to substitute intelligence here) are controlled, income appears to enhance fertility. See Deborah S. Freedman, "The Relation of Economic Status to Fertility," *American Economic Review,* 53 (June, 1963), pp. 414-426; and Otis Dudley Duncan, "Marital Fertility as a Career Contingency" (12 April, 1966), unpublished manuscript.

46. There is no contradiction here between the "regression toward the mean" and the principles of "evolution." For instance, students frequently ask how the giraffe's long neck could have evolved to its present length if the offspring of exceptionally long-necked adults were shorter on the average than their parents. The answer is simple: the regression is toward the mean of the *total* adult population of the *parent's* generation. If, owing to some selection process, the exceptionally long-necked giraffes were more fertile than others in a particular generation, a new mean would be established for the offspring. Although the necks of the offspring of the long-necked adults would tend to be somewhat shorter than those of their parents, the new mean would nevertheless be higher than the old mean of the previous generation. Also, the necks of some offspring would probably exceed the longest of any adults in the previous generation. Evolution is slow, but certain. Furthermore, there is no reason to believe that the genotypes associated with intelligence do not operate in precisely the same manner.

47. It is possible that part of the regression toward the parental mean, in terms of the phenotypic range of intelligence, is also due to environmental variation, but only to the extent that the trait-relevant environments are not positively related to the genotypes associated with higher intelligence.

48. Certainly it may be hard to convince any intelligent middle-class parent that the chances are that the "native" intelligence of any one of his children probably is less than his own and perhaps not much better than average. The parent, however, is hardly in any position to judge.

49. Barrington Moore, Jr., *Political Power and Social Theory,* Cambridge: Harvard University Press, 1958, p. 163. Also, note one of Lipset and Bendix' concluding remarks:

    There can be no doubt . . . that the discrepancy between the distribution of intelligence in a given generation of youth and the distribution of social positions in the parental generation is a major dynamic factor affecting mobility in all societies in which educational achievement or other qualities associated with intelligence play an important role in status placement.

See Seymour Martin Lipset and Reinhard Bendix, *Social Mobility in Industrial Society,* Berkeley: University of California Press, 1959, p. 236. The statement, of course, would be more applicable to my argument if we were to substitute the term "intelligence" for "social position" in their phrase "the distribution of social position in the parental generation." To the extent that "qualities associated with intelligence play an important role in status placement," the substitution does not appear inappropriate.

50. Readers may recognize the obvious linkages here to the large body of literature on deviant behavior that has developed from Merton's "means-ends" paradigm. However, no research, to my knowledge, has attempted to explain either lower- or middle-class delinquency in terms of Merton's model by taking into account the discrepancies between the parents' and child's intelligence, or even between the parents' status and the child's intelligence. Are not these discrepancies probably just as much a part of the "structure of opportunities" as those that are culturally prescribed? John A. Clausen has raised a similar question in a recent (1966) unpublished ms., "The Organism and Socialization."

51. James Bryant Conant, *Slums and Suburbs,* New York: McGraw-Hill, 1961, p. 144.

52. David Goslin, Roberta R. Epstein, and Barbara A. Hallock, *The Use of Standardized Tests in Elementary Schools,* New York: Russell Sage Foundation, 1965; Orville G. Brim, Jr., David A. Goslin, David C. Glass, and Isadore Goldberg, *The Use of Standardized Ability Tests in American Secondary Schools and Their Impact on Students, Teachers, and Administrators,* New York: Russell Sage Foundation, 1965; and Orville G. Brim, John Neulinger, and David C. Glass, *Experience and Attitudes of American Adults Concerning Standardized Intelligence Tests,* New York: Russell Sage Foundation, 1965.

53. Clearly one of the largest and most comprehensive (and costly) surveys ever undertaken in this country, I nevertheless have found few sociologists who are any more than vaguely familiar with Project Talent. Baseline data were gathered in 1960 when a two-day battery of psychometric tests and biographical questionnaires were administered to nearly one-half million 9th through 12th

grade students in roughly 1,000 public and private high schools. Adding to these auspicious beginnings, long-range plans call for a series of follow-ups one, five, ten, and 25 years after each class graduates. It is, indeed, unfortunate that sociologists have not been more aware of this "data bank" and its potential applications. Among the several monographs published to date, segments of the above findings were reported in John C. Flanagan, John T. Dailey, Marion F. Shaycoft, David B. Orr, and Isadore Goldberg, *Studies of the American High School,* Pittsburgh: Project Talent Office, University of Pittsburgh, 1962.

54. Owing to their similarity in content, some schools, no doubt, use the PSAT, too, as a "practice" test for the SAT (the most widely used of the college admissions tests) which generally is administered the following year.

55. Inadvertently, the National Merit examinations also are being used as a basis for ranking both the high schools and the colleges. On the one hand, the absolute number of "finalists" has become a mark of considerable prestige among the secondary schools. (Sociologists and others should be warned, however, that the "prestige" accorded any school on the basis of these awards is not relative to any set of national norms, but relative only to other schools in the same state. The reason is that different cutting points on the National Merit examination are established for each state in order that equal proportions of their enrollments are eligible for the awards.) In regard to ranking the colleges, the attempts of most researchers have been frustrated either because different colleges use tests which are not always comparable (even though the College Board's SAT holds a pre-eminent position) or because published scores are not available. In lieu of these problems, Astin recently devised a "selectivity" scale for nearly all colleges in the country on the basis of the "choices" indicated by the National Merit semi-finalists and recipients of its Letter of Commendation in 1961. (The "choices" are regularly obtained since the size of the stipends they award are based, in part, on tuition fees.) An influential work, particularly among the pre-college guidance counselors, is Alexander W. Astin's *Who Goes Where to College?,* Chicago: Science Research Associates, 1965.

56. In regard to the National Merit examination, see the *Annual Report,* National Merit Scholarship Corporation, Evanston, Illinois, 1965. Of special interest to sociologists should be the fact that NMSC has been obtaining socioeconomic data along with the administration of its tests. As these data continue to accumulate in the years ahead, we should have a rather good historical record of the interrelationships between ability, social class, and college choice.

57. See note 55.

58. For example, whereas in 1955–56 only 1,229 students who entered 130 colleges from 104 secondary schools took the Advanced Placement examinations, the respective figures for the year 1963–64 were 28,874 students entering 885 colleges from 2,086 secondary schools. See, *College Decisions on Advanced Placement* (a CEEB Research and Development Report), Princeton, N.J.: Educational Testing Service, January, 1966. Associated with the growth of both the APP and the NMSC, the number of honors programs in American colleges tripled between 1957 and 1965. Moreover, it generally is recognized that the basic fundamental of honors work is an insistence that each student fulfills his potential; see, Joseph W. Cohen, ed., *The Superior Student in American Higher Education,* New York: McGraw-Hill, 1966.

59. I have only touched on some of the highlights in the development of educational testing. Furthermore, I have neglected almost entirely any consideration of the use of standardized tests in business, government, and the military service. For a more complete overview of ability testing today, see Goslin, *The Search . . . , op. cit.*

60. It is doubtful that the controversy related to "invasion of privacy" that came to a climax in the 1965 congressional investigations will, in the long run, stem the tide. Nor is it likely that the more recent threats of legislative intervention, even if they materialize, will significantly restrict the actual administration of tests or other instruments in our schools. For a thorough review of testing and public policy, see the November 1965 issue of the *American Psychologist,* 20.

61. It is difficult to generalize about the privately endowed schools, some of which flourish (expand) in response to the demand for

higher education, while others hold their size and simply become more selective. Most of the state-supported schools, on the other hand, appear to respond in very much the same two ways, both of which require these schools to employ some device, like testing, to decide who should go to college and where. One factor is that before a state-supported school can expand its faculties or facilities it must provide the state legislature with appropriate evidence that its requests are warranted. Since "projected" enrollment figures probably are not too convincing, the only acceptable evidence seems to be the number of "qualified" students who had to be "turned away last year." There is, therefore, a constant lag between demand and supply which requires the colleges and universities to make an "acceptable" choice between who should and who should not be admitted. The second factor involves the differentiation in purpose and function of different colleges within the same system. As increasingly larger segments of the college-age population look for some form of post-secondary education, the major institutions of higher learning seek to protect their standards in new ways. At one time, it was acceptable to admit anyone and then simply "weed out" the poorer students during the freshman year. However, the method probably was devastating to both the students and the faculty. Nevertheless, as enrollment pressures have mounted, the practice could not survive without making administrative "monsters" out of the schools whose immense size already creates far too many problems. As a result, we are witnessing the proliferation of the junior and community colleges and the remarkable transformation of the teacher's colleges and "cow" colleges into liberal arts schools, accompanied by the establishment of formal criteria, including testing, to decide who goes where.

62. I have in mind here something in addition to the more traditional separation of the "academic" and "vocational" schools. In a number of large public school systems around the country, entire units have been set aside in recent years not just as preparatory schools for the college aspirants but specifically for the "gifted" students. For example, the student bodies at Bronx High and Hunter High in New York City, as a whole, rank in the 99th percentile on standardized tests. See, Cohen, *op. cit.,* p. 226. (Roughly three-fourths of them, too, win recognition on the National Merit examinations.)

63. To a limited extent, selection into the private sector occurs across all socioeconomic classes. For example, highly selective independent schools like Exeter and Andover "sponsor" a "fair" number of students from lower socioeconomic backgrounds. (The author, along with Richard E. Peterson at Educational Testing Service, is currently involved in a study of all living alumni from the "Exeter community.")

64. In addition to the eventual determination of college plans, formal evaluation takes place early in order to identify both children with very exceptional abilities and those who, owing to their deficiencies, require special attention. See David A. Goslin, *The School in Contemporary Society,* Chicago: Scott, Foresman, 1965, p. 110.

65. Vernon, *op. cit.,* p. 143. It should not be surprising that, in a physiological sense, the brain centers associated with intelligence are slow to mature. This seems to be characteristic of the human child as a whole. With a few exceptions, like the sucking reflex, the human infant enters the world exceedingly ill-equipped, for some time, to deal with his environment. See Weston LaBarre, *The Human Animal,* Chicago: University of Chicago Press, 1960.

66. These forces also appear to explain what kinds of tests are administered and when. Whereas IQ tests generally are used as screening devices in the early primary grades when specific kinds of learnings or experiences are perhaps somewhat less likely to invalidate the tests, aptitude and achievement tests are more often used later when little or no pretense is required concerning what the tests actually measure. Psychometricians freely admit that achievement tests are bound to specific kinds of subject matter, since this is the purpose for which they are designed. Most, however, are much less willing to admit that their aptitude tests are "subject-bound."

67. Burton R. Clark, *The Open Door College,* New York: McGraw-Hill, 1960. In addition to the "cooling-out" function that Clark so aptly describes, another "compensatory factor" involves the repeated use of standardized tests of various forms and at various points in the educational process. For example, some observers believe that the PSAT (see note 54) which usually is administered during the eleventh grade and just a year before the college entrance examinations is an unnecessary proliferation of tests. They

argue that the accumulated scores from other tests administered throughout the primary and earlier secondary grades are sufficient both to predict how well most students will do on the entrance examinations and to "narrow down" the choice of a college. Certainly, for predicting how any "group" of students will do, these observers probably are correct. On the other hand, multiple testing nevertheless makes it possible to detect and take into account the uneven inter-individual patterns of intellectual growth (plus changes in motivation and error terms) that undoubtedly affect every student's test performance.

68. Ralph H. Turner, "Sponsored and Contest Mobility in the School System," *American Sociological Review,* 25 (December, 1960), pp. 855-867. While Turner suggested that the United States has been moving toward a system of sponsorship, the proliferation of testing in this country really was just beginning when he wrote this article.

69. In my judgment, the best complete text on the sociology of education that has appeared so far, Corwin's discussion of the "talent hunt" nevertheless strongly reflects the biases of the environmentalists. See Ronald G. Corwin, *A Sociology of Education,* New York: Appleton-Century-Crofts, 1965, especially pp. 191-207.

70. Brian Jackson, *Streaming: An Education System in Miniature,* London: Routledge and Kegan Paul, 1964.

71. Not only is our own "field of vision" restricted by pre-college attrition and selection, it is largely limited to the relatively narrow range of aptitudes that characterizes the student body at any particular college or university. Owing to increased diversity and selectivity in higher education, the inter-institutional differences in intelligence (or scholastic aptitude) are usually about as great as, and often greater than, the intra-institutional differences. See, for example, T. R. McConnell and Paul Heist, "The Diverse College Student Population," in Nevitt Sanford, ed., *The American College,* New York: Wiley, 1962, pp. 225-252.

72. For example, see David Glass, *Social Mobility in Britain,* London: Routledge and Paul, 1954; Natalie Rogoff, *Recent Trends in Occupational Mobility,* Glencoe: Free Press, 1953; Gosta Carlson, *Social Mobility and Class Structure,* Lund: C. W. K. Gleerup, 1958;

and W. Lloyd Warner et al., *The American Federal Executive,* New Haven: Yale University Press, 1963.

73. In a quite different context, Rogoff discusses this problem with reference to predicting the rate of college-going among persons of different ability and different class origins. See Natalie Rogoff, "American Public Schools and Equality of Opportunity," in A. H. Halsey, Jean Floud, and C. Arnold Anderson, eds., *Education, Economy, and Society,* New York: The Free Press of Glencoe, 1961, pp. 140-147.

74. A. H. Halsey, "Genetics, Social Structure and Intelligence," *The British Journal of Sociology,* 9 (1958), pp. 15-28; and J. L. Gray and P. Moshinski, *The Nation's Intelligence,* Watts, 1936.

75. Both Burt and Anderson have used models that estimate the difference between the expected and observed frequencies of mobility on the basis of class differentials in the distribution of intelligence. In Anderson's model, for example, "ideal" mobility is defined as the perfect correspondence between intelligence and occupation whereby the degree of departure from the "ideal" is taken essentially as an index of inequality. While this model avoids the trappings I described above, it has the objectionable weakness of making no allowance for culturally-induced variations in measured intelligence. Anderson recognizes this objection and freely admits that the model is simply based on the "efficiency" goal of economic theory. Burt, on the other hand, is much less sensitive to the problem. See Cyril Burt, "Intelligence and . . . ," *op. cit.;* and C. Arnold Anderson, James C. Brown, and Mary Jean Bowman, "Intelligence and Occupational Mobility," *Journal of Political Economy,* 60 (June, 1952), pp. 218-239.

76. For an excellent discussion of the kinds of "discriminatory imparities" I have in mind, and with which we *should* be concerned, see C. Arnold Anderson and Philip J. Foster, "Discrimination and Inequality in Education," *Sociology of Education,* 38 (Fall, 1964), pp. 1-18.

77. Some authors, such as Cattell, believe they have come close to one, however. In Cattell's case, a "culture-fair" test has been developed that purportedly measures an individual's ability to adapt to new

situations and in which his "crystallized" (learned) skills are of no particular advantage. In comparison with other measures of intelligence, he maintains that this test has about twice the normal standard deviation, is more stable with changes in age or culture, and, therefore, is more biologically determined. See Raymond B. Cattell, "Theory of Fluid and Crystallized Intelligence," *Journal of Educational Psychology,* 54 (February, 1963), pp. 1-22. My own belief is that the most significant advances in this regard are not likely to be made by psychometricians. The idea of a completely "culture-fair" test is, as others have suggested, probably absurd. Rather, any new breakthroughs are likely to be the work of biochemists. Since the chemical structure of the genetic "alphabet" has been discovered, one can only agree with Dobzhansky that "the day may not be too far away when the sequences of the genetic 'letters' in the various genes in man and in other organisms may become known." See Theodosius Dobzhansky, *Heredity and the Nature of Man,* New York: Harcourt, Brace, and World, 1964, p. 37.

78. Aaron V. Cicourel and John I. Kitsuse, *The Educational Decision-Makers,* New York: Bobbs-Merrill, 1963. Nearly as extraordinary as the recent growth in educational testing has been the establishment of formal guidance programs. It is estimated that nearly 90 percent of the nation's high schools have trained guidance counselors today, as against about 36 percent in 1955. See Project Talent, *op. cit.,* chapter 3, p. 38.

79. In addition to his work previously noted, see David A. Goslin, "The Social Consequences of Predictive Testing in Education" (1965), and "The Social Impact of Testing in Guidance" (1966), both unpublished manuscripts.

80. William H. Sewell and J. Michael Armer, "Neighborhood Context and College Plans," *American Sociological Review,* 31 (April, 1966), pp. 159-168; and commentaries by Ralph Turner, Richard Boyle, John Michael, William Sewell, and J. Michael Armer, *American Sociological Review,* 31 (October, 1966), pp. 698-707. In Sewell's defense, I wish to note that the evidence from several longitudinal studies clearly indicates that during adolescence the intra-individual (test-retest) correlations on IQ across this age range are very stable under normal conditions, i.e., about as high

as the reliabilities of the tests employed. Moreover, the small *gains* that can be isolated from one point in time to another appear to be correlated with the earlier level of ability, despite either Benjamin Bloom's (*op. cit.*) or Robert Thorndike's ("Intellectual Status and Intellectual Growth," *Journal of Educational Psychology,* 57 [June, 1966], pp. 121-127) pessimism. The best evidence is yet unpublished but comes from the "Growth Study" at Educational Testing Service. In a closely administered eight-year program involving the repeated use of the same or equated tests in 23 schools, early results indicate that the children who show rapid intellectual growth over one period of time tend to be the same children who had a prior history of rapid growth. If this is the case, there perhaps is less room for "neighborhood context" than previously thought, especially after socioeconomic (family) status has been partialed out. Nevertheless, the possibility still exists that a *small* part of the association between intelligence and neighborhood can be attributed to the effect of the latter on the former, and, to this extent, Sewell and Armer probably should have been more cautious.

81. Otis Dudley Duncan, "Intelligence and Achievement: Preliminary Results" (22 April, 1966) and "Intelligence and Achievement: Further Calculations" (20 July, 1966), both unpublished manuscribe. Duncan's data tentatively suggest that while intelligence has a significant effect on occupation apart from its correlation with family background, all or nearly all of its effects operate within the context of the school system and therefore do not add to the "explained" variance in occupation that can be accounted for by family background and education alone.

82. Bernard Farber, "Social Class and Intelligence," *Social Forces,* 44 (December, 1965), pp. 215-225.

83. Ralph Linton, *The Study of Man,* New York: Appleton-Century, 1936.

84. For example, see David Matza, *Delinquency and Drift,* New York: Wiley, 1964.

E.
Genetics and Ethnic Characteristics

Ethnic groups frequently have characteristics which are largely genetic in origin. The three papers in this section show disease patterns in Jews, fertility levels in Welsh, and behavioral differences between Chinese-American and European-American newborns. Each succeeding paper probably incorporates more components of environmental influences which affect the resulting data.

The genetic and disease patterns data presented by Sheba show both similarities and dissimilarities of Jewish subgroups who have been nearly isolated for hundreds of years. For example, color blindness is fairly constant in all subgroups of Jews. On the other hand, Kurdistan Jews show very high frequencies for Beta-thalassemia where none is shown in the European/U.S. subgroup. There are also remarkable differences in many gene frequencies between Ashkenazy Jews and non-Ashkenazy Jews even though the geographical separation between the two groups is probably only 25-50 generations. It might represent a fascinating study to relate gene frequencies, language differences, and cultural characteristics with the widely separated population of the Jews.

Ashley in the introduction of his paper suggests that Welsh and non-Welsh differ in some genetic characters. He further finds that the larger the component of Welsh in a population sample, the lower is the fertility index in married women. In the paper he examines a number of other variables such as socio-economic status, the differences of populations by names, age at marriage, age at first delivery, and urban and rural distribution. All of these are rejected as adequate explanations of lower fertility for Welsh women. This leads him to conclude that the lower fertility pattern among Welsh women is in all likelihood genetic in origin.

Overall fertility is a more complex problem to study than gene frequencies for diseases as seen in the Sheba paper. Fertility problems may involve many genetic pathways any one of which may decrease fertility. In addition, environmental complexities intrude at many stages of development. The geneticist must show that significant environmental factors are not operating to produce the result in that specific sample. Then after the process of elimination he/she can suggest that genetic factors are very likely the causal factors.

The paper by the Freedmans indicates that Chinese-American newborns were generally calmer and steadier than those of European-American newborns. The researchers appear to have controlled many significant co-variables. One is inclined to believe that the ethnic differences are based on genetic contributions to behavioral patterns. The paper is intriguing and inevitably should lead the reader to consider the possible effects of uterine existence as a contributing environmental factor.

Additional Readings

Goldsby, Richard A., 1971. *Race and Races.* The Macmillan Company, New York. A paper on sickle cell disease should have been included in this volume but unfortunately none could be found to satisfy the selective conditions described in the preface. Accordingly, it is recommended that Goldsby's book be consulted for a good introduction on sickle cell problems.

Dellaportas, G. J., 1971. Birth weight, ethnic origin, and perinatal mortality. *Social Biology,* **18:**158-163. The data on reproductive problems indicates that a predominantly French-Canadian Quebec Province sample has lower birth weight, increased perinatal risk of death, and other differences when compared to a sample of predominantly English-Canadian individuals from the rest of Canada. In many respects this is similar to Ashley's results on fertility.

Scheinfeld, Amram, 1965. *Your Heredity and Environment.* J. P. Lippincott Company. An older but still useful popular introduction to human heredity. See especially the fascinating comparison between two individuals: one American by birth but culturally Chinese since infancy and the other racially Chinese but culturally American. The author of the book was responsible for bringing together these two culturally transposed individuals. (Pages 624, 625)

Lee, J. Warren, 1955. Tongue-folding and tongue-rolling in an American Negro population sample. *Journal of Heredity,* **46:**289-291. The study indicates that there are substantial differences between Afro-American and Chinese populations in lingual movements.

Sharma, G. C., 1968. Convergent evolution in the tribes of Bastar. *American Journal of Physical Anthropology,* **28:**113-118. Investigator finds that four different tribes in Central India which are genetically and socially isolated have similar genetic traits. It is suggested that similarities between the tribes is due to a common geographical environment for hundreds of generations.

King, James C., 1971. *The Biology of Race.* Harcourt Brace Jovanovich, Inc. The section on "Traditional misconceptions about human variation" is very useful for this section on Genetic and Ethnic Characteristics

Questions

Do you think parental desire in the planned-for number of children is the exclusive factor in overall fertility? The major factor?

It has been shown that greater differences in genetic compositions are generally correlated with longer separation in time between populations. What can you say about probable genetic "closeness" as provided in the data by Sheba?

The behavioral differences between the Chinese-American and European-American newborns appear established. Would you suspect that the differences are (a) all genetic, (b) genetic and environmental, or (c) all environmental? Explain.

11. Gene-Frequencies in Jews

CHAIM SHEBA

Sir,—During the past 25 years I have been investigating gene-frequencies in exiles returning to Israel. It seems to me that some genetic studies of conditions in Jews have been based on an oversimplified concept—that of a simple division of Jews into Ashkenazim and non-Ashkenazim. It may be that I was partly responsible for this oversimplification when, in 1945, I showed that constitutional haemolytic anaemia is found only in non-Ashkenazim.[1] I should like to present a short account of the genetic distribution of Jews, and of the gene frequencies of some conditions found in the different groups.

The Ashkenazim were the Jews who lived in Europe, and later in the United States and Israel. They spoke Yiddish—a sort of mittelhoch-deutsch-cockney written in Hebrew characters—and could also be identified by the setting of the liturgy in their prayer books.

The non-Ashkenazim come under the group heading, Sepharadim (sepharad=sunset). The common language of the Sepharadim is the Ladino—ancient Castilian Spanish written in Hebrew characters of the kind used in Southern France by Rashi, the Bible commentator of the eleventh century. These Jews provided all Jewry with the bitter lemon, the "ethrog," for the feast of the tabernacle. Splinter groups of Jews lived as far away as Cochin, on the southwestern tip of India, and the Yemen, on the southwestern tip of Arabia. Another isolated group of Jews lived in Hadramaut (Court of Death), 6 weeks' camel distance from Aden. There were also, of course, large Jewish congregations in Morocco, Algiers, Tunis, Libya, and Egypt, and very small remnants in Pakeen, in Upper Gallilee, and Sichem, in Samaria. There was a flourishing independent community in Babylon (now Iraq), with satellites in Iran, Kurdistan, the Caucasus mountains, Afghanistan, Syria, the Lebanon, and Asia Minor. The Balkan Jews of Greece, Bulgaria, Yugoslavia, and Italy really belonged to the Sephardic group (with the possible exception of Stephen Dunn's 200 families). The accompanying figure shows a scheme of the distribution of the various groups, designed for me by Dr. Israel Ashkenazy, of Tel Aviv University.

Reprinted by permission from *The Lancet,* 1970, June 6, 1230-1231.

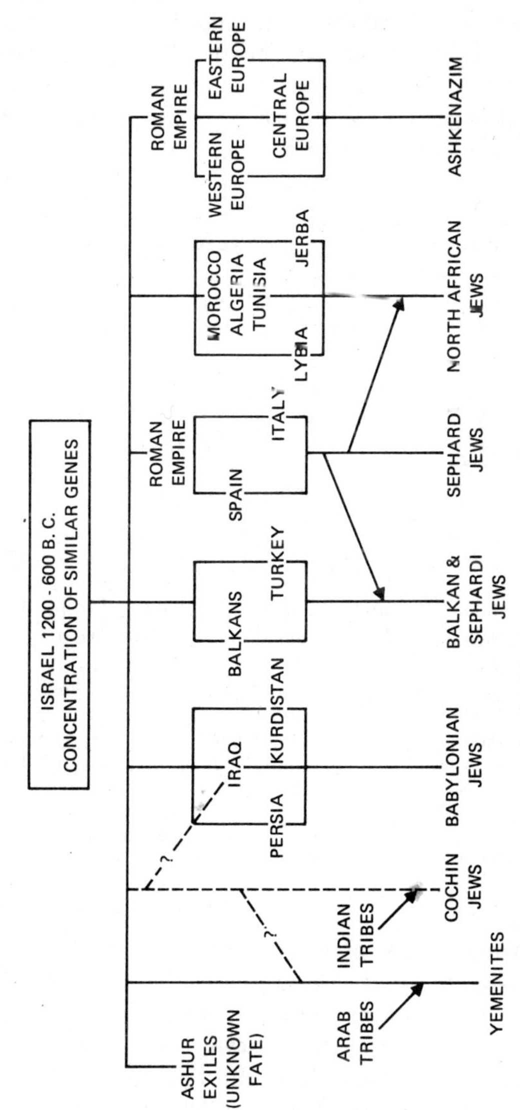

Fig. 1 Ethnic distribution of Jews from Israel between 1200 B.C. and A.D. 1970.

Now a brief account of some of the gene frequencies. Rough estimates for some of the most extensively studied conditions are shown in the accompanying table. In addition, the following conditions are almost exclusively found in Ashkenazy as opposed to non-Ashkenazy Jews: Tay-Sachs disease, Gaucher's disease, Buerger's disease, pentosuria, and dysautonomia. (Congenital pyloric stenosis is also about three times more prevalent than among non-Ashkenazim.)

Rough estimates of gene-frequencies in seven Israeli communities

	Place of origin						
	Morocco Algeria Tunesia	Libya	Yemen	Iraq	Kurdistan	Iran	Europe USA
Population (thousands):	415	60	140	210	30	80	1100
Conditions*:							
G. -6P.D. def.	0.006	0.01	0.05	0.25	0.58	0.15	0.003
β -thalassaemia	0.005	(?)0	Rare	0.01	0.10	0.015	0
F.M.F.	0.02	0.04	0	0.015	9 cases	0	10 cases
D.J.S.	9 cases	0	0	5 cases	4 cases	64 cases	7 cases
P.K.U.	0.005	0	0.019	0.009	0	0.015	0
Atyp. pseudochol.	0.003	–	0.036	0.05	–	(?) 0.05	0.017
Colour blindness	0.06	0.06	0.04	0.05	0.05	0.06	0.09

*Gene frequencies unless otherwise stated. G. -6P.D. def. = glucose-6-phosphate dehydrogenase deficiency. F.M.F. = familial Mediterranean fever. D.J.S. = Dubin-Johnson syndrome. P.K.U. = phenylketonuria. Atyp. pseudochol. = atypical pseudocholinesterase.

The following conditions are *almost* exclusively found in non-Ashkenazim: glycogen-storage disease, pituitary dwarfism (with high growth-hormone levels), vitamin-B_{12} malabsorption, and Wolman's disease (familial xanthomatosis). Libyan Jews, with the highest gene-frequency of familial Mediterranean fever, must have been a special isolate, as were the Jews of Ispahan (Iran), with the highest frequency of Dubin-Johnson syndrome.

Notes

1. Schieber, C. *Lancet,* 1945, ii, 851.

These data were assembled from material in the Tel-Hashomer Institute of Human Genetics, and with the help of Avinoam Adam, Israel Ashkenazy, Mariassa Bat-Miriam, Bathsheba Bonné, Bernie Cohen, Joseph Gafni, Eliahu Gilon, Harry Heller, Nehammah Kossover, Amira Mani, Baruch Padeh, Mordechai Pras, Bracha Ramot, Yecheskiel Samra, Uri Seligsohn, Mordechai Shani, Ezra Sohar. I am especially grateful to Aryeh Szeinberg, of Tel-Hashomer and Elisabeth Goldschmidt, of Jerusalem.

12. Welshness and Fertility

DAVID J. B. ASHLEY

The population of Wales can be divided into two parts, the Welsh and the non-Welsh, who differ in some genetic characters (Ashley and Davies, 1966a) and in their susceptibility to a number of common diseases (Ashley and Davies, 1966b; Ashley, 1966, 1967a, b, c). Both moieties of this population share the environment of the Principality of Wales and have similar occupations and similar social status (Ashley and Davies, 1966a) but can be separated on the basis of two parameters, the ability to speak the Welsh language and the possession of a Welsh surname.

The present study is directed to an investigation of the differential fertility of the Welsh and non-Welsh people of Wales.

I: GEOGRAPHICAL DIFFERENCES

Wales can be divided clearly into three zones according to the frequency with which the Welsh language is used (Registrar General, 1962a). In the western counties of Anglesey, Caernarvon, Merioneth, Cardigan, and Carmarthen, which comprise the area of the Glyndwr revolt of 1400–1415 (Rees, 1951), more than 65% of the people are Welsh speaking; in the counties of Brecon, Denbigh, Flint, Glamorgan, Montgomery, and Pembroke, and in the towns of Merthyr Tydfil and Swansea between 15 and 40% of the people are Welsh speaking, and in the remainder of the country, the counties of Monmouth and Radnor and the towns of Cardiff and Newport less than 10% of the people claim to speak Welsh.

A mean fertility index was calculated for these three zones for the years 1958 to 1963 inclusive. The number of live and stillbirths per year per 1000 women between the ages of 15 and 44 was determined for each area, using the data of the Registrar General for England and Wales (1960, 1961, 1962a, b, 1963, 1964a, 1965). In the area of high Welsh speaking the mean fertility index was 79; in the area of intermediate Welsh speaking it was 89, and in the area of low Welsh speaking it was 96 (Table 1).

Reprinted by permission from *Journal of Medical Genetics,* 1969, Volume **6,** 180.

The proportion of women who were married at the time of the 1961 census varies in different areas: in the high Welsh area it was 64% and in the other two areas was rather higher, 69% (Registrar General, 1964b). The fertility index has therefore been recalculated as the number of live and stillbirths per year per 1000 married women between the ages of 15 and 44 (Table 1). The frequencies of illegitimate births, 5.73% in the high Welsh area, 5.3% in the intermediate area, and 5.75% in the low Welsh area were similar, and have been disregarded for this purpose of comparison.

The fertility index for married women shows a maximum in the low Welsh areas and a minimum in the high Welsh areas.

In England and Wales as a whole there are differences in fertility in the different urban and rural areas. In towns of over 100,000 inhabitants the fertility index per 1000 married women is 134, in towns of between 50 and 100,000 people it is 140, in smaller urban areas it is 124, and in rural areas it is 134. The three areas of Wales have different proportions of town and country; in the low Welsh area half the people live in the two big towns of Cardiff and Newport, while in the high Welsh area over 90% live in rural areas. The expected fertility index for married women for these areas was calculated on the basis of the proportion of the population living in the different urban and rural areas, and was compared with the observed index (Table 2). The ratio between the two was calculated, and shows a significant gradient between the high, intermediate, and low areas.

Table 1. Fertility index in Wales

Welsh speaking	Female population 15–44	Live and stillbirths	Fertility index	% Women married	Corrected fertility index births/ 1000 married women/yr.
High	78,872	37,509	79	64	123
Intermediate	288,341	153,434	89	69	129
Low	142,020	80,327	96	69	139

Table 2. Observed and expected fertility index for married women

Welsh speaking	Observed	Expected	Observed/ expected
High	123	133	92.5
Intermediate	129	132	98
Low	139	133	105

Fertility is related to socio-economic status, but analysis of the fertility index of the counties and county boroughs of Wales did not show a significant correlation with socio-economic status nor did a similar calculation on the fertility indexes in the Registrar General's standard Regions.

II: DIFFERENCES BY NAMES

Data were available on the obstetric history and the married and maiden names of 593 women over the age of 25 years who were delivered in the Maternity department at Morriston Hospital during the year 1962. These patients were divided into three age-groups 25-29, 30-34, and 35 years and over, and classified according to their married and maiden names. The number of patients, the total number of pregnancies, and the mean number of pregnancies were calculated for each group (Table 3).

Within each group the greatest mean number of pregnancies was seen in the patients whose married and maiden names were both non-Welsh; in the youngest and the oldest groups the lowest mean number of pregnancies was seen in the patients whose married and maiden names were both Welsh. Names were regarded as Welsh if they were included in the list of Welsh names given in the previous papers (Ashley and Davies, 1966a, b).

The numbers of pregnancies expected in each group, by age, were extracted and summed in Table 4. There was a deficiency of pregnancies in the patients with Welsh married names and also in the patients with Welsh maiden names. This deficiency was statistically significant in the case of the maiden names. There was a significant excess of pregnancies in the group of patients whose married and maiden names were both non-Welsh.

Table 3. Numbers of pregnancies according to Welsh or non-Welsh names

Age (yr.)	Married name	Maiden name	No.	No. of pregnancies	Mean No. of pregnancies
25–29	Welsh	Welsh	72	164	2.28
	Welsh	Non-Welsh	59	137	2.32
	Non-Welsh	Welsh	56	138	2.46
	Non-Welsh	Non-Welsh	64	173	2.70
			251	612	2.44
30–34	Welsh	Welsh	49	131	2.80
	Welsh	Non-Welsh	41	110	2.68
	Non-Welsh	Welsh	43	111	2.58
	Non-Welsh	Non-Welsh	50	150	3.00
			183	508	2.78
35–	Welsh	Welsh	43	128	2.97
	Welsh	Non-Welsh	35	125	3.57
	Non-Welsh	Welsh	42	128	3.05
	Non-Welsh	Non-Welsh	39	143	3.67
			159	524	3.30

A possible reason for the difference in fertility is that, for cultural reasons, the Welsh might tend to marry later than the non-Welsh or to practise family limitation more effectively. No data are presently available on the latter point. The age at first pregnancy could, however, be examined and compared with the surname in a total of 342 women delivered of their first babies at Morriston Hospital during the year 1962. twenty-two of these women were single, they were generally younger than the married women and have been excluded from the analysis.

The mean age at first delivery in the women with Welsh married and maiden names was slightly greater than in the women with non-Welsh names. The differences were small and no significant difference was observed. The slightly higher age at first delivery in the women whose married and maiden names were dissimilar may indicate a delay in choosing a spouse from some other geographical area.

Table 4. Number of pregnancies in women with Welsh and non-Welsh married and maiden names

		No. observed	No. expected	Observed/ expected
Married name	Welsh	801	827.2	97
	Non-Welsh	843	818.6	103
Maiden name	Welsh	806	848.5	95
	Non-Welsh	838	797.3	105
Both married and maiden names	Welsh	429	453.8	95
	Non-Welsh	466	423.9	110
One name Welsh, one name Non-Welsh		749	767.1	98
Married name Welsh, maiden name Non-Welsh		372	373.4	99.5
Married name Non-Welsh, maiden name Welsh		377	394.7	96

Table 5. Mean age at first delivery

Married name	Maiden name	Mean age (yr.)
Welsh	Welsh	24.5
Welsh	Non-Welsh	25.4
Non-Welsh	Welsh	25.2
Non-Welsh	Non-Welsh	24.0
Married name	Welsh	24.9
	Non-Welsh	24.6
Maiden name	Welsh	24.8
	Non-Welsh	24.7

DISCUSSION

The fertility of the Welsh women has been shown to be lower than that of their non-Welsh sisters by the two methods of analysis which have previously been used in studies of disease incidence. Fewer babies are born in the western, Welsh speaking counties than in the more heavily anglicized south eastern part of the country; this difference is not related to differences in social class nor to the different urban-rural distribution in the different parts of Wales, and is probably not due to a later age of marriage in Welsh women. Study of women delivered of children in Morriston Hospital, which lies in one of the intermediate Welsh speaking areas of the Principality, shows that the average number of pregnancies is lower in women with Welsh married or maiden names than it is in women with non-Welsh names. Welsh women married to non-Welsh men and non-Welsh women married to Welsh men occupy an intermediate position.

It is suggested that the reason for this lower fertility among Welsh women is genetic in origin. The Welsh people form, with the peoples of the other areas of the 'Celtic Fringe,' the last remaining elements of the earliest peoples of Britain who have been pushed westward by succeeding waves of immigrants. They have retained their language, which is learned and spoken by very few immigrants, and still show quite a high degree of assortative mating among themselves (Ashley and Davies, 1966a). This leads to some degree of genetic isolation and to the formation of a different gene pool in each of the two populations of Wales, the Welsh and the non-Welsh. If the difference in fertility was only seen in the geographical study it could be postulated that it was due to environmental differences between the three different areas, but when it is combined with a difference between women coming from the same environment to a single hospital whose only difference is in their names the inference that the difference is genetic becomes much stronger.

The data of Table 4 suggest that the difference in fertility is greater in the wives than in the husbands, though this does not reach the level of statistical significance. The nature of the difference, the way in which a difference in genotype is expressed as a difference in fertility is uncertain. We have shown (Ashley and Davies, 1966a) that there are differences in the A B O and Rh blood group distributions of the Welsh and non-Welsh people of the Swansea area, and it is possible that minor differences in the frequency of genes such as these may interact to affect over-all fertility (*Brit. Med. J.,* 1954) or even that differences in the leucocyte and transplantation antigens (van Rood, van Leeuwen, and Bruning, 1967) may be responsible.

Cultural differences between the Welsh and non-Welsh could be responsible for the difference in fertility. Analysis of the local data, however, showed no significant difference in the age at first delivery. Data on the use of contraceptive methods in the two groups were not available, and deliberate family limitation in one group remains a possibility.

SUMMARY

Studies of the fertility of women in the three areas of Wales in which Welsh speaking is common, intermediate, and rare, and of the obstetric experience of women with Welsh and non-Welsh married and maiden names show that there is a lower fertility among the women of Wales.

It is suggested that this difference is genetic in origin and is related to the different gene pools of the Welsh and non-Welsh components of the population of Wales.

This work was carried out with the aid of a research grant from the Welsh Hospital Board. The substance of the paper was presented at a meeting of the Society for the Study of Human Biology in March 1968.

References

Ashley, D. J. B. (1966). Observations on the epidemiology of prostatic hyperplasia in Wales. *Brit. J. Urol.* **38,** 567.

——— (1967a). Coronary artery disease in Wales. *J. Med. Genet.,* **4,** 277.

——— (1967b). Diabetes in Wales. *ibid.,* **4,** 274.

——— (1967c). Vascular disease of the central nervous system in Wales. *ibid.,* **4,** 280.

———, and Davies, H. D. (1966a). The use of the surname as a genetic marker in Wales. *ibid.,* **3,** 203.

———, and ——— (1966b). Gastric cancer in Wales. *Gut,* **7,** 542.

Brit. Med. J. (1954). Annotation. Fitness, fertility, and blood group. **1,** 1197.

Rees, W. (1951). *An Historical Atlas of Wales.* Faber and Faber, London.

Registrar General (1960). *Statistical Review of England and Wales for 1958. Part I. Tables Medical.* H.M.S.O., London.

——— (1961). *ibid.,* for 1959.

______ (1962a). *Census 1961 Wales. Report on Welsh Speaking Population.* H.M.S.O., London.

______ (1962b). *Statistical Review of England and Wales for 1960. Part I. Tables Medical.* H.M.S.O., London.

______ (1963). *ibid.,* for 1961.

______ (1964a). *ibid.,* for 1962.

______ (1964b). *Census 1961 England and Wales. Age, Marital Condition and General Tables.* H.M.S.O., London.

______ (1965). *Statistical Review of England and Wales for 1963. Part I. Tables Medical.* H.M.S.O., London.

van Rood, J. J., van Leeuwen, A., and Bruning, J. W. (1967). The relevance of leucocyte antigens for allogeneic renal transplantation. Symposium on tissue and organ transplantation. *J. Clin. Path.,* **20,** Suppl., 504.

13. Behavioural Differences between Chinese-American and European-American Newborns

D. G. FREEDMAN AND NINA CHINN FREEDMAN

In the course of an investigation of newborn behaviour among widely distributed ethnic groups, some interesting preliminary results have appeared.

Twenty-four Chinese-American and twenty-four European-American newborns were examined while still in the nursery. The Orientals were largely of Cantonese background and the Caucasians largely of middle-European background. All families were middle class and the bulk were members of a pre-paid health plan (Kaiser Hospital, Table 1). Table 1 summarizes the potentially important co-variables other than ethnic group, and we see that none could have accounted for the ethnic differences presented here.

The behaviour scales we used were modifications of a procedure developed by Brazelton; a full description is available from the authors and will be published soon. Briefly, they consist of twenty-five general behavioural items rated 1-9, and fifteen standard neurological signs, usually used to screen for neural damage, rated 0-3. The twenty-five general items may be arranged into five categories as follows: (1) temperament—seven items; (2) sensory development—four items; (3) central nervous system maturity—six items; (4) motor development—five items; (5) social interest and response—three items.

All testing was done during September and October 1968. Each test session lasted between 30 and 40 min. Testing was performed in the newborn nursery by N. C. F. as D. G. F. watched, and scoring was done immediately afterwards in a room next to the nursery. Apart from a reliability sample of four infants, scoring depended on verbal agreement between the authors. Four arbitrarily selected infants formed the reliability sample and of the 160 items involved the authors were over 1 point apart in only three instances; all scales reported below yielded reliability coefficients of 0.912 or better, with an average reliability of 0.969.

A multivariate analysis of variance indicated that, on the basis of

Reprinted by permission from *Nature,* 1969, Volume 224, No. 5225, p. 1227.

Table 1. Comparisons of potential co-variables

	Chinese-American	European-American
Mean age and range in hours (n.S.)	32.75 (7–75)	33.27 (5–72)
Initial state (rated 1–6, from deep sleep to very alert) means (N.S.)	3.58	2.79
Distribution of sexes*	11 male, 13 female	11 male, 13 female
Mean birth weight in grams† ($P = 0.05$)	3,194.33	3,447.91
Mean Apgar‡ rating at 5 min after birth (N.S.)	8.86	9.00
Mean hours of labour (N.S.)	6.08	5.77
Medication during labour§	16 received systemic drugs 8 received only local anaesthetic or none	13 received systemic drugs 11 received only local anaesthetic or none
Mean age of mothers (N.S.)	26.70	26.66
Mean number of previous pregnancies (N.S.)	1.83	2.41
Hospital	16, Kaiser Hospital, San Francisco 5, Chinese Hospital, S.F. 3, U.C. Medical Center, S.F.	20, Kaiser Hospital, S.F. 4, Lying-in, Chicago

*There was no significant interaction between race and sex.
†When weight is treated as a co-variable it does not affect ethnic differences.
‡A rating of viability; based on heart rate, colour, respiration, tonus, and crying. Optimal score is 10.
§Although systemic drugs significantly lowered Apgar ($P = 0.02$), automatic walk ($P = 0.02$) and tonic deviation of the head ($P = 0.02$), statistical treatment indicates that these drug differences did not affect ethnic differences.

total performance, the two groups were decidedly different ($P = 0.008$). Further analysis indicated that the main loading came from the group items measuring temperament and which seemed to tap excitability/imperturbability ($P = 0.001$). While the following discussion is based on mean ethnic differences on the distinguishing items, it should be emphasized that there was substantial overlap in range on all scales between the Chinese and Caucasian infants.

The European-American infants had a greater tendency to move back and forth between states of contentment and upset (lability of states) ($P = 0.01$), and they showed more facial and bodily reddening ($P = 0.005$), probably as a consequence. The Chinese-American infants were scored on the calmer and steadier side of these items. In an item called defensive movements, the tester placed a loosely woven cloth firmly over the supine baby's face for a few seconds. While the typical European-American infant immediately struggled to remove the cloth by swiping with his hands and turning his face, the typical Chinese-American infant lay impassively, exhibiting few overt motor responses ($P = 0.0001$). Similarly, when placed in the prone position, the Chinese infants frequently lay as placed, with face flat against the bedding, whereas the Caucasian infants either turned the face to one side or lifted the head ($P = 0.02$). Inasmuch as there was no difference between the groups in the ability to hold the head steady in the upright position ($P = 0.91$, pull to sit), this maintenance of the face in the bedding is taken as a further example of relative imperturbability or ready accommodation to external changes. In an apparently related item, rate of habituation, a pen light was repeatedly shone on the infants' eyes, and the number of blinks counted until the infant no longer reacted (shut-offs). The Chinese infants tended to habituate more readily ($P = 0.06$).

There were no significant differences in amount of crying, and when picked up and consoled both groups tended to stop crying. The Chinese infants were, however, often dramatically immediate in their cessation of crying when picked up and spoken to, and therefore drew extremely high ratings in consolability ($P = 0.007$). The Chinese infants also tended to stop crying sooner, without soothing (self-quieting ability, $P = 0.06$).

To summarize, the majority of items which differentiated the two groups fell into the category of temperament. The Chinese-American newborns tended to be less changeable, less perturbable, tended to habituate more readily, and tended to calm themselves or to be consoled more readily when upset. In other areas (sensory development, central nervous system maturity, motor development, social responsivity) the two groups were essentially equal.

We thank Dr. T. Berry Brazelton for training us in these methods, the cooperating hospitals and staff, and Murray Edelman and David Truslow for statistical help and advice.

F. Origins of Human Matings: Genetic and Social Issues

We are always interested in the genetic and cultural backgrounds of human matings and the consequences of these matings. As pointed out in a previous section, assortative matings are nonrandom because there has been a preferential selection of the spouse with a particular characteristic. Eckland discussed how couples are frequently drawn to one another by their intellectual abilities.

Another kind of mating involves inbreeding. This is a mating of closely related individuals generally having one or more immediate common ancestors. This type of mating can involve second cousins, first cousins, even aunt-nephew, uncle-niece, and brother-sister matings. A survey of the currently available elementary texts on genetics indicates that comprehensive materials on inbred matings are generally provided. Accordingly, a paper on this subject is omitted from this volume.

Most elementary texts do not provide much material on the founder principle or founder effect. This frequently has considerable bearing on inbreeding or consanguineous matings. The founder principle operates within certain colonies or small groups of individuals who possess, somewhat fortuitously, distinct genetic endowments of one or a few founder individuals. Once this founding group is established, the initial genetic resources of the founding colony will tend to persist. In a sense it becomes an inbred colony. The small colony with its peculiar genetic composition may increase and become a major population of mankind which still retains the distinct genetic endowments of the original colony. For example, Tay-Sachs disease is predominantly found in Jews, whose ancestors lived in northeastern Poland and southern Lithuania. In New York City it is estimated that one in three have this awesome recessive gene. When Tay-Sachs disease occurs in non-Jewish persons, it is usually shown that the affected children are the result of an inbred or consanguineous mating. In one of the papers in this section, Livingston deals with the founder effect and deleterious genes. The paper provides additional insight into the genetic bases of Tay-Sachs disease in Jewish populations as well as the sickle cell disease of black populations.

A mating between two individuals who have no family ties in the immediate past and who may be separated by ethnic or racial differences may be called an outcrossing. Occasionally, in the scientific literature, the term outbreeding is used. However, strictly speaking, this should be reserved for genetically unrelated organisms whereas human ethnic or racial groups are regarded as evolutionarily related. Two papers in this section deal with outcrossing.

Weyl in his paper on plantation slavery surveys the results of racial mixture of the cDe chromosome combination. He finds that tests on this chromosome in Afro-Americans from a metropolitan area reveal the admixture to be higher than those tested in rural areas. This, Weyl claims, is historically linked with miscegenation, emancipation, and then migration.

In a commentary on the Weyl paper and in particular the section entitled Virginia and the internal slave trade, Allen (1971 *Social Biology* 18:104) has written "genetic factors doubtless influence some of the behavior in which treatment was based but the historical data seems too scanty to indicate any net effects of selection other than the elimination of extremes. Present geographic distribution of eminent blacks must mainly reflect differential opportunities."

The paper by Bresler relates fetal loss, defined as abortions plus stillbirths, in an F_1 level to different numbers of national origin three generations back. It was found that with increasing national origin, or presumed heterogeneity, the greater was the fetal loss. Over the years genetic counselors have been opposed to inbred matings because of the potential adverse genetic consequences. If these findings on outcrossings are correct, it represents an additional caution whereby matings of individuals from markedly different genetic groups may also have adverse affects on reproduction.

Additional Readings

Pinto-Cisternas, J., C. Salinas, C. Campusano, H. Figueroa, and B. Lazo, 1971. Preliminary migration data on a population of Valparaiso, Chile. *Social Biology,* **18:**305-310. Individuals whose parents or grandparents were not born in Valparaiso had significantly different gene frequencies in the ABO blood group system and socio-economic index than those individuals whose parents were born in Valparaiso. Accordingly, migrants and nonmigrants showed genetic differences.

Bhattacharya, D. K., 1969. A study of ABO, Rh-Hr and MN blood groups of the Anglo-Indians of India. *Human Biology,* **41:**115-124. Anglo-Indians show a greater contribution from the Europeans in both Rh and MN systems than they show in the ABO system. The genetic contribution of the British Isles to Anglo-Indians is approximately 39.5%.

Johnston, Francis E., 1970. A phenotypic assortative mating among the Peruvian Cashinahua. *Social Biology,* **17:**37-42. The social regulation of marriage common to all human societies narrows the range of potential spouses in a way which interferes with random mating. At the genotypic level this nonrandomness is referred to as inbreeding. At the phenotypic level nonrandom mating is referred to as assortative mating. Although spouse selection may have a genetic basis, most previous studies have shown that assortative mating has been for education, intelligence and religion but some of them have been shown to work for height and weight as well. This study shows that there is some positive assortative mating for height and weight between husband and wife.

Eckland, Bruce K., 1970. New mating boundaries in education. *Social Biology,* **17:**269-277. An analysis shows that assortative mating for intelligence is a built-in feature of the educational selection system.

Alstrom, Carl Henry, 1970. Some problems in population genetics. *Journal of Medical Genetics,* **7:**289-293. A study of a Swedish sub-population of isolate type in 1800 and 1825 showed that 55% of the second generation was derived from 26% of the prior generation.

Questions

Can you think of any voluntary or enforced migrations where the group leaving the country probably had a different set of behavioral (presumably genetic as well) characteristics than those who remained behind?

Under what conditions in an urbanized community could a founder effect take place? Is it possible?

Assuming that interethnic matings do in fact increase fetal loss, do you believe that this may be a major factor in reducing the completed family size?

14. The Founder Effect and Deleterious Genes

FRANK B. LIVINGSTONE

Many distinctive human populations are characterized by the presence of one or more lethal or severely deleterious genes in frequencies which would be defined as polymorphic according to Ford's ('40) famous definition. The particular genetic disorder, however, varies. The Old Order Amish of Lancaster County, Pennsylvania have a gene frequency of 0.07 for the recessive Ellis-van Creveld syndrome, while the Amish as a whole have a frequency of about 0.05 of the recessive cartilage-hair hypoplasia syndrome (McKusick et al., '64). Many of the tri-racial isolates of Eastern United States also have a high frequency of a deleterious gene (Witkop et al., '66). Although such populations are frequently defined by religious or ethnic criteria, there are others not so defined. Several island populations in the Aland archipelago have a gene frequency of greater than 0.1 for von Willebrand's disease (Eriksson, '61), and the Boer population of South Africa and some populations of Northern Sweden have frequencies of porphyria much greater than those of other populations (Dean, '63; Waldenstrom and Haeger-Aronsen, '67). However, these conditions are dominant and do not have the very severe effects of other hereditary disorders found in high frequencies. On the other hand the population of the Chicoutimi District of Quebec has recently been found to have a gene frequency of about 0.02 for tyrosinemia, which is a lethal recessive (Laberge and Dallaire, '67).

In most of these cases the population in question has undergone a rapid increase in recent years, and the question arises as to whether this rapid expansion and the original small size of the isolate could account for the high frequency of the deleterious gene. Such an explanation by the founder effect seems obviously to apply to most of the cases cited above, but the founder effect may well be a more general explanation of human gene frequency differences. It is now becoming apparent that the major populations of mankind vary significantly in their frequencies of deleterious genes and that many large populations such as Eastern European Jews have high frequencies of deleterious genes which are found in low frequencies in other populations (McKusick, '66). There have

Reprinted by permission from *American Journal of Physical Anthropology,* 1969, Volume 30, 55.

been many attempts to determine how such genes could be polymorphic, for example, Anderson et al. ('67) and Knudson et al. ('67) have discussed cystic fibrosis and Myrianthopoulos and Aronson ('66), Tay-Sachs disease. The purpose of this paper is to attempt to determine the extent to which the founder effect can cause high frequencies of deleterious genes with various models of population expansion.

The occurrence which initiated this research is the gene for sickle cell hemoglobin in the Brandywine isolate of Southeast Maryland. At present the sickle cell gene frequency in this isolate is about 0.1 (Rucknagel, '64). The high frequencies of this gene in many parts of Africa, India, and the Middle East are now well-accepted as being due to a relative resistance of the sickle cell heterozygote to falciparum malaria. The high frequency in the Brandywine isolate may have a similar explanation, but the surrounding Negro population does not have such a high frequency. And although the endemicity of falciparum malaria in Southeast Maryland in the last century is not known in any detail, it would not appear to have been great enough to explain the high sickle cell frequency in the Brandywine isolate. The isolate also has many other deleterious genes in high frequency (Witkop et al., '66).

The Brandywine isolate seems to have had its beginning in the early Eighteenth Century when laws were passed to prohibit co-habitation and marriage among races, which prior to then were presumably frequent or at least known. Up to 1720 there were several prosecutions under these laws of individuals with surnames currently present in the isolate (Harte, '63). Harte ('63) has maintained that the Brandywine isolate is derived from these illegal unions, and Witkop et al. ('66) show that the most common surname came from such a union. In 1790 the first United States Census recorded 190 persons with the group's surnames as "other free people," and since then over 90% of the recorded marriages have been endogamous or between individuals with surnames within the group (Harte, '59). According to Harte ('59) there are six "core" surnames which have been associated with the group since its founding and comprise 66% of the population and another ten surnames which entered the group after the Civil War, but Witkop et al. ('66) list seven core surnames and eight marginal ones. The total population of the isolate is now estimated to be 5,128 (Witkop et al., '66), and the statistics do indicate rapid, if erratic, growth (Gilbert, '45; Harte, '63).

In order to simulate gene dynamics the population has been assumed to have doubled itself in early generations, and then after slower growth to have approached a doubling in recent generations. The simulation was run for 10 generations with the following numbers in succeeding generations: 20, 40, 80, 160, 320, 640, 664, 728, 856, 1112. This approximates the early demographic history of the Brandywine isolate, but the isolate is much larger today. However, gene frequency change in later generations with a large population is very small.

The simulation model randomly selects two parents from the initial population which has been assumed to have either one or two sickle cell heterozygotes among the founders. A family size is randomly determined, the offspring generated and then selected out or stored with no compensation for those not surviving (a copy of the program is available on request). Since the population is increasing rapidly, the family size distribution approximates that recorded by Roberts ('65) for a population in Tanzania which has about 4.0 surviving offspring per female. The founder population can actually vary in size, however. The size of the offspring generation is the number which is set; but with an average of 4.0 offspring per marriage and 40 offspring, the founder population would be expected to consist of ten marriages or 20 individuals.

Figure 1 shows the distribution of the deleterious gene frequencies after ten generations for two sets of 50 runs each with different initial conditions and different fitnesses for the genotypes. With a gene frequency of 0.05, which is comparable to having two sickle cell heterozygotes among the founders, the gene is present at a frequency of greater than 0.04 in almost 40% of the populations, while for a starting gene frequency of 0.025 or one founder with the sickle cell trait, 16% of the populations have the gene at a frequency of greater than 0.04. With two founders there were runs which resulted in a gene frequency as high as that of the sickle cell gene in the Brandywine isolate, but with one founder there were none as high. However, there were many frequencies close to it, so that such an outcome is possible if not probable. Hence there seems to be no necessity to postulate a selective advantage for the sickle cell in the Brandywine isolate. It should be pointed out that this simulation and further ones assume the population is closed. Gene flow from other populations would tend to decrease the frequency of the deleterious gene, but if most of the population's expansion is due to natural increase, then the founder effect would be most important.

In order to determine whether such high frequencies could occur in a population with a greater number of founders, a similar program was run with 40 and 80 founders. The results are shown in figure 2. These

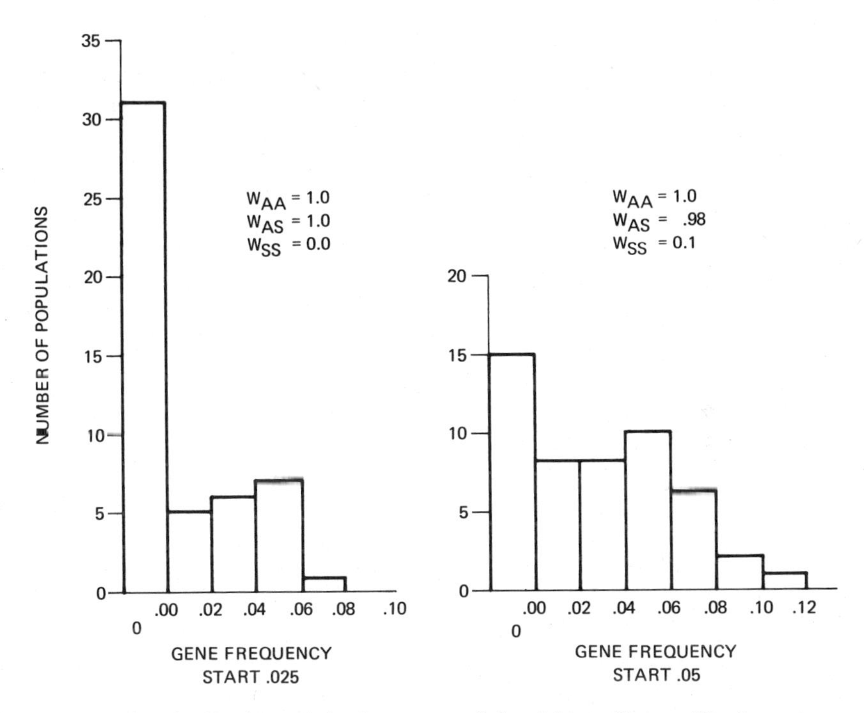

Fig. 1 The distributions of the frequency of the sickle cell gene (S) after ten generations of expansion for two sets of 50 runs each with different initial gene frequencies and different fitnesses for the genotypes (W's). Note: the populations in which the S gene has been completely eliminated are separated in the left column from those in which it is still present.

runs were started with one carrier of the lethal gene and the population doubled itself for five generations, so that it ended with 1,280 and 2,560 individuals, respectively. Although the lethal gene is not present in high frequencies in as many populations, it is still present in about 5% in a frequency greater than 0.04. The fact that populations begun with a few founders should have such high frequencies of lethal genes seems to indicate that they can contribute to the problem of the genetic load. According to Morton ('60), the average individual has the equivalent of four recessive lethals in the heterozygous state. For a population with 40 founders this would imply over 100 lethal or deleterious genes among the founders, so that several would be expected to attain high frequencies. The fact that the number of deleterious genes in small populations started by a few founders seems to average much less may indicate a lower genetic load, but the Brandywine isolate, the Amish, and the Eastern European Jews do have several deleterious genes in high frequencies. In any case, it seems to be a possible way to study the problem.

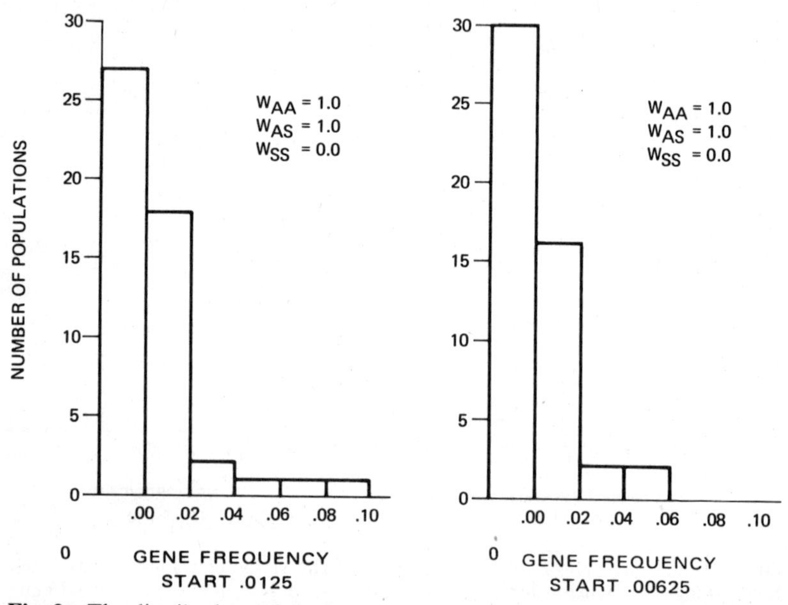

Fig. 2 The distributions of the frequency of a lethal recessive gene (S) after five generations of expansion for two sets of 50 runs each with different initial gene frequencies which represent one carrier of the lethal gene in a founder population of 40(0.0125) and one carrier in a population of 80(0.00625). Note: the populations in which the lethal gene has been completely eliminated are separated in the left column from those in which it is still present.

The most recent population expansion which seems to have increased the frequency of a lethal gene is the peopling of the Saguenay River and Lake St. John region by French Canadians. Settlement of the Upper Saguenay did not begin until the 1830's, and Chicoutimi was founded in 1840 by 220 individuals from La Malbaie, 66 from Eboulements, and 37 from Baie St. Paul (Buies, 1896). In 1861 the population of Chicoutimi was 10,478 and in 1871 it rose to 17,483. Much of this increase was undoubtedly due to immigration, but given the enormous rate of increase of the French Canadian population as a whole (Henripin,

'54), the expansion of the population of Chicoutimi was due to a great extent to natural increase. Under these conditions a lethal gene frequency of 0.02 would not be unlikely and seems to agree with our simulation model. The fact that the entire French Canadian population stems from about 10,000 original settlers (Henripin, '54) may have led to this population having its own set of lethal genes.

The high frequencies of deleterious genes in the Eastern European Jewish populations of Lithuania and Eastern Poland may have a similar explanation, although there is disagreement about this possibility (McKusick, '66). Myrianthopoulos and Aronson ('66) do not consider such an explanation likely for the Tay-Sachs gene. Instead they propose a slight selective advantage for the heterozygote. They have to postulate the operation of this selective advantage for 50 generations which is longer than the population has been there. It is also much longer than the factor thought to confer the selective advantage, typhoid fever in the ghettos (Aronson, '64), seems to have been present as a serious disease. The Tay-Sachs gene attains its highest frequencies in the Jewish populations of Southern Lithuania and Northeast Poland, which were founded in the Twelfth Century after the Crusades led to the persecution of the Jews in Germany. Although the Jewish settlements in Lithuania were founded by refugees from the west, according to Herzog ('65) they preceded by two or three centuries the Jewish settlements in Mazovia to the west in Poland. Thus, these colonies were isolated for some time and were actually expanding to the west into Northern Poland when the Jews were expelled from Lithuania in 1495. Most moved to adjacent territories but then moved back to Lithuania in 1503. Hence, the population history of these Jewish groups seems to be one of expansion from a few founders. In any case by the time of the flowering of Eastern Jewish culture in the Sixteenth Century, the population was very large and continued to expand up to the Twentieth Century.

When a population size of 1,000 or more is attained, the change in gene frequency is approximated by deterministic equations. For a lethal recessive the frequency after n generations is:

$$q_n = \frac{q^o}{1 + nq_o},$$

where q_o is the initial gene frequency. A lethal gene in a large population is thus eliminated at a very slow rate, particularly when it occurs in a very low frequency. If the Tay-Sachs gene increased to 0.05 in the early generations before the population became large, in the approximately 30 generations since then, the gene would have decreased to

$$\frac{0.05}{1 + 30(0.05)} = 0.02,$$

which is about the frequency today in Eastern European Jews.

The fact that lethal genes are eliminated at such a slow rate in large populations would make it possible for them to have "polymorphic" frequencies long after the original expansion. Since most of the world's populations have expanded rapidly in the last 1,000 years, much of the variability in the frequencies of lethal genes (or non-lethals for that matter) could be a consequence of the original expansions of the major populations. As an example, it is suggested that this effect may explain the high frequencies of cystic fibrosis in the populations of Europe, which range around a gene frequency of 0.02. For the non-Caucasian populations on Hawaii Wright and Morton ('68) have estimated the gene frequency for cystic fibrosis to be 0.003, which presumably is close to the equilibrium frequency due to a balance of selection and mutation. Given this frequency, over 50% of a set of founder populations of size 100 would be expected to have a carrier of this lethal. With the sudden expansion of such a set of founder populations it seems possible that such a lethal could attain a frequency of 0.05 for the entire population. A more precise mathematical expression of the problem seems possible and could perhaps yield a solution.

Literature Cited

Anderson, C. M., J. Allan and P. G. Johansen 1967 Comments on the possible existence and nature of a heterozygote advantage in cystic fibrosis. In: Cystic Fibrosis. E. Rossi and E. Stoll, eds. Bibliotheca Paediatrica, No. 86. S. Karger, New York, pp. 381-387.

Aronson, S. M. 1964 Epidemiology. In: Tay-Sachs' Disease. B. W. Volk, ed. Grune and Stratton, New York, pp. 118-154.

Buies, A. 1896 Le Saguenay et le Bassin du Lac Saint-Jean. Léger Brousseau, Québec, 3rd edition.

Dean, G. 1963 The prevalence of the porphyrias. So. Afr. J. Lab. Clin. Med., **9:** 145-151.

Eriksson, A. W. 1961 Eine neue Blutersippe mit v. Willebrand-Jürgens' scher Krankheit (erbliche Thrompathie) auf Åland (Finnland). Acta Genet. Med. Gemell., **10:** 157-180.

Ford, E. B. 1940 Polymorphism and taxonomy. In: The New Systematics. J. S. Huxley, ed. Clarendon Press, Oxford, pp. 493-513.

Gilbert, W. H. 1945 The wesorts of Maryland; an outcasted group. Jour. Wash. Acad. Sci., **35:** 237-246.

Harte, T. J. 1959 Trends in mate selection in a tri-racial isolate. Social Forces, **37:** 215-221.

______ 1963 Social origins of the Brandywine population. Phylon, **24:** 369-378.

Henripin, J. 1954 La Population Canadienne au Début du XVIIIe Siècle. Institut national d'études démographiques, Paris, Cahier No. 22.

Herzog, M. I. 1965 The Yiddish Language in Northern Poland: Its Geography and History. Indiana University Research Center in Anthropology, Folklore and Linguistics, Pub. No. 37.

Knudson, A. G., L. Wayne and W. Y. Hallett 1967 On the selective advantage of cystic fibrosis heterozygotes. Amer. J. Hum. Genet., **19:** 388-392.

Laberge, C., and L. Dallaire 1967 Genetic aspects of tyrosinemia in the Chicoutimi Region. Canad. Med. Assoc. J., **97:** 1099-1100.

McKusick, V. 1966 Clinical genetics at a population level. The ethnicity of disease in the United States. Alabama J. Med. Sci., **3:** 408-424.

McKusick, V., J. A. Hostetler, J. A. Egeland and R. Eldridge 1964 The distribution of certain genes in the Old Order Amish. Cold Spring Harbor Symp. Quant. Biol., **29:** 99-114.

Morton, N. E. 1960 The mutational load due to detrimental genes in man. Amer. J. Hum. Genet., **12:** 348-364.

Myrianthopoulos, N. C., and S. M. Aronson 1966 Population dynamics of Tay-Sachs Disease. I. Reproductive fitness and selection. Amer. J. Hum. Genet., **18:** 313-327.

Roberts, D. F. 1965 Assumption and fact in anthropological genetics. Jour. Roy. Anthrop. Inst., **95:** 87-103.

Rucknagel, D. L. 1964 The Gene for Sickle Cell Hemoglobin in the Wesorts. Thesis, University of Michigan, Ann Arbor.

Waldenström, J., and B. Haeger-Aronsen 1967 The porphyrias: a genetic problem. In: Progress in Medical Genetics. Volume V. A. G. Steinberg and A. G. Bearn, eds. Grune and Stratton, New York, pp. 58-101.

Witkop, C. J., C. J. MacLean, P. J. Schmidt and J. L. Henry 1966 Medical and dental findings in the Brandywine isolate. Alabama J. Med. Sci., **3:** 382-403.

15. Some Genetic Aspects of Plantation Slavery

NATHANIEL WEYL

Plantation slavery was necessarily characterized by variations in mortality and life expectancy between the skilled and the unskilled, the craftsmen and the field hands, the house servants and the plantation gangs, the workers on cotton farms and on sugar estates, and the bondsmen and the free. The processes affecting birth rates and mortality within the Negro population were considerably more complex than abolitionist literature suggests, varied a great deal in relation to time and place, and were by no means unidirectional.

The estimates most commonly used of the white genetic component in the American Negro exaggerate the true state of affairs because they are based on urban Northern black populations, in which interracial gene flow was abnormally high. Differential mortality rates favored cotton and tobacco workers as against rice and sugar workers, house servants as against plantation hands, craftsmen as against unskilled laborers. This tended to give Negroes with white blood higher rates of natural increase.

Manumission of slaves, which was highly selective for white genes and which was regarded, by masters and servants alike, as a signal mark of favor, generally operated in the opposite direction. Demographic conditions conspired to decrease the rate of population growth among free Negroes far below that among slaves. This condition prevailed in the United States both because the medical care which masters gave their slaves from reasons of self-interest was markedly better than that which the free Negroes obtained for themselves and because the emancipated black population was unwanted by white society, despised, impoverished, and socially degraded. In areas such as the West Indies where the mulatoo and free Negro populations enjoyed a more stable social status (one superior to that of the slaves), this difference may not have prevailed. If so, the comparative survival rates of free Negroes and black-white crosses would be materially different in the Antilles than was the case in the United States.

Reprinted by permission from *Perspectives in Biology and Medicine,* Summer 1970, 618-625.

GEOGRAPHY OF RACE MIXTURE

The best single serological index of the percentage of Negro genes is the R^0 chromosome combination in the Rhesus blood groups, which is cDe in the Fisher notation. Since the bulk of slaves came from West Africa, an unweighted mean was taken of the six samples from the region reported by Mourant [1, p. 394]. The average R^0 was 58.5 percent.[1] By comparison, Glass and Li found an R^0 percentage of 43.8 for their Baltimore Negroes, as compared with 2.8 percent for white Americans [2]. Thus, if the Baltimore sample were representative of the nation, American Negroes would have an average of 26.4 percent non-Negro genes. This, however, is not the case. A 1963 study by Workman, Blumberg, and Cooper of the serology of Negroes in Evans and Bullock counties in Georgia reported a mean R^0 component of 53.5 percent, indicating a racial admixture of 9.0 percent [3]. A study of James Island, South Carolina, blacks yielded an R^0 of 51.7 percent, or an admixture of 12.2 percent [4], and a 1958 examination of Charleston Negroes by W. S. Pollitzer revealed an R^0 factor of 56.5 percent, which would indicate only 3.6 percent white ancestry [5].

The differences between the indicated values for *m,* the coefficient of racial admixture, between the Baltimore and the rural Southern samples reflect the fact that miscegenation, emancipation, and migration have been historically linked. The Negro population of the Carolina-Georgia Sea Islands, of Charleston and of the northern Georgia counties studied by Workman et al. consisted to a large extent of Negroes from Angola, called *Gullahs,* who worked the large-scale coastal cotton and rice plantations, had few social contacts with their white masters, and were therefore largely isolated from interracial gene flow. The Baltimore sample of Glass and Li *per contra* consisted disproportionately of the descendants of free Negroes and of artisan and house slaves. Wherever the institution of chattel slavery prevails, emancipation is positively and high correlated with miscegenation. Since the Southern states regarded both mulattoes and free Negroes as security risks and took energetic measures to extrude them, the Negro population of the North was always more racially mixed than that of the South. Thus, according to 1860 census figures, 31 percent of the black population of the Northern states was mulatto, but only 12 percent of the black population of the slave states. Within the South, 50 percent of the free Negroes, but only 7 percent of the slaves, were classified as mulatto [6, p. 12].

A serological map of the Negro population of the United States, based on R^0 percentages, would provide the factual foundation for a realistic analysis of migration and gene flow. Unfortunately, this does not exist. Provisionally, we might estimate the Caucasian genetic component in the present American Negro population as midway between the Baltimore and the rural Southern figures, or at 16 percent.

A county-by-county analysis of the relationship between mental test scores and R^0 percentages would shed light on the hypothesis suggested by Jensen [7] that negative correlations exist between Negro genes and psychometric intelligence. Efforts to obtain this information from the Department of the Army (AFQT scores) and from the HEW custodians of the Coleman Report were unavailing, despite the pledge made in the latter report to "make all the data gathered by this survey available to research workers"[8, p. 2].

PROCESSES AFFECTING MORTALITY

Complaint about the mortality in the sugar areas was general and international in scope. Sir Charles Lyell, an English visitor who abhorred slavery, wrote that in Louisiana "the duration of life for a sugar mill hand does not exceed seven years" [9, p. 35]. In Cuba, Klein, a modern authority, concluded that sugar work was "exacting and grueling. Hours were long, work hard, *mayorales,* or overseers, extremely cruel, and mortality quite high by the island's standards" [10, p. 150]. Only raw *bozal* males, that is, blacks freshly introduced from Africa, were used and, during the busy season, the Negroes were allowed only five hours of sleep. "Before the introduction of the steam engine and the example of a milder treatment of the Negro by foreign residents," an American physician wrote in 1844, "the annual loss by death was fully ten percent, including, however, new slaves, many of whom died from the change of climate" [11, p. 153]. Despite improvement in the life expectancy of black cane hands after 1830, Cuban urban slaves seem "to have been corrected more by the threat of being shipped to the [sugar] plantations than of being whipped in town"[10, p. 155].

Captain John Gabriel Stedman, whose *Narrative of an Expedition to Surinam*[2] is a neglected masterpiece, spent five years as a mercenary officer, putting down a Negro revolt in the Dutch colony. Stedman lived openly with his mulatto slave mistress, acknowledged their child, and became an effective critic of the barbarities of chattel slavery under Netherlands rule. Among the many subjects he investigated was the anatomy of plantation slavery, considered chiefly in terms of differential death rates. He estimated that there were 80,000 slaves in the colony. Of these, some 30,000 (artisans, house workers, and specialized estate workers) "live better than the common people of England" (!) another 30,000 "are kept in idleness and do not work in the fields"; whereas, the remaining 20,000 are "among the most miserable wretches on earth; and are worked, starved, insulted, and flogged to death" [12, p. 393].

Stedman noted that, although 2,500 Negroes were imported into Surinam from Africa annually, the slave population of the colony failed to increase. Since "each Negro has a wife or several if he wishes," there should have been no barrier to reproduction. Asserting that it was on the 20,000 field hands that "chiefly falls the dreadful lot of untimely mortality," Stedman inferred a life expectancy for a plantation gang worker of about ten years [12, p. 373].

Bryan Edwards, the classic historian of the West Indies, asserted that slavery in the British islands was mild, despite the admitted fact that a large annual import of African Negroes was needed to prevent the black population from dwindling. He argued that birth rates were held down by a paucity of female slaves, by polygamy among the Negroes, and by the "extreme licentiousness and profligacy of manners in most of their women, with frequent abortions and barrenness" [13, p. 428]. During the Colonial period, the estimated net annual decrease of the slave stock ranged from 4 to 7 percent yearly [14, p. 472]. Males were imported by preference, since sugar planting was heavy work. High food costs and the nearness of West Africa made importation more economical than breeding.

In the United States, the lower Mississippi, with its sugar estates, inspired the slaves with well-founded dread. After the transition from French to American control, "it became notorious that slaves were overworked, underfed and brutally treated to such a degree that the rate of mortality exceeded the birth rate" [14, p. 520]. A progressive improvement in living conditions followed the abolition of the African slave trade in 1808. Nevertheless, the 1850 census showed a slave mortality rate of 24.0 per thousand in Louisiana as against 16.4 per thousand in the nation as a whole. While the 1849 cholera epidemic accentuated this difference, the fact that the Negro population of the area had fewer old people operated in the contrary direction [15, p. 96].

The rice plantations were also insalubrious. White labor could not survive there due to malaria, from which the West African Negroes were partially protected by heterozygous sickling genes. Nonetheless, work in the rice fields caused severe rheumatism, pleurisy, and other pulmonary diseases [16, p. 170].

VIRGINIA AND THE INTERNAL SLAVE TRADE

The primary source of slaves for the expansion of the new cotton belt of 1830–1860 and the development of the sugar lands of Louisiana was the depleted soil of the Eastern Seaboard. The charge that Virginia bred slaves as an industry is exaggerated, for this would imply that she selected male studs and enforced sexual abstinence when prices were low. The Commonwealth did, however, export almost 300,000 Negroes south and west during 1830–1860, accounting for almost the total increase of her servile population.

These slaves were not chosen at random. They were characterized by Ulrich B. Phillips, a defender of the Old South, as "the indolent, the unruly and those under suspicion" [17, p. 192]; by the contemporary Southern journalist D. R. Hundley (1840) as "the most refractory and brutal of the slave population" [18]; and by Kenneth M. Stampp, a historian with an impeccably abolitionist approach to the peculiar institution, as including "slaves accused of felonies . . . slaves sentenced to be transported beyond the limits of the United States," and, in one instance, twenty-four slaves "convicted of murder, burglary, rape, arson, manslaughter, and attempted insurrection" [19, p. 258]. After the Nat Turner rebellion, Virginia swarmed with slave traders ready to buy up Negro suspects. Selling south was deemed a means of separating "a vicious Negro" from more tractable ones.

In Virginia, slaves of good character would be advertised for sale provided they were not sent out of the state and provided their families were not broken up [17, p. 191]. There were always two prices for slaves in Virginia, a domestic price and a foreign price for shipment south which was generally one-half to two-thirds higher [17, p. 192].

On the other hand, runaway slaves were often punished by being sold south, and Flanders concluded that, in the state of Georgia at least, "in the great majority of cases the runaways were mulattoes, the characteristic phrases of 'copper-colored' or 'inclined to be a little yellow' recurring throughout the advertisements" [16, p. 216].

Depending on the preconceptions and prejudices of historians, the selection of Virginia Negroes for sale south has been treated as a process for eliminating criminal and vicious elements or as one designed to reward docility and eliminate potential leaders of revolt. In either case, the selection process should have had genetic repercussions of a more or less permanent character. In an effort to check this, a count was made of the listings on every sixth page of *Who's Who in Colored America* (1950 edition), and the count was then multiplied by six. These totals of eminent Negroes were then compared with the 1910 census figures of Negro population of the Southern states. In Alabama, Mississippi, Texas, and Louisiana—four deep South recipient states of the domestic slave trade—the average number of distinguished Negroes per million of Negro population was 158, as against a coefficient of 411 per million in Virginia, the chief slave-exporting state. This seemed to establish a *prima facie* case for the view that the sale of slaves south and west was negatively selective.

FREE NEGROES AND SLAVES

Between 1850 and 1860, according to the census, the slave population increased by 28.8 percent and the free Negro population by only 12.3 percent. The figures for 1840–1850, although somewhat vitiated by the annexation of Texas with its large slave population, tell approximately the same story. The increases in both slave and free Negro numbers during 1850–1860 were almost entirely due to natural increase. Slave increments from the illegal African trade were probably more than offset by manumissions, escapes, and colonization of Liberia. The free Negro population probably gained more from emancipation than it lost from emigration and "passing for white," yet its increase was only 40 percent that of the slaves.

Contemporary records in those Northern cities with good birth and death statistics reveal a fairly uniform pattern of high mortality among the free Negroes. In Boston, during 1855–1859 inclusive, Negro deaths exceeded Negro births by a bit less than two to one; in Philadelphia in 1860, deaths among the free Negroes were more than double births. Providence reported a Negro death rate of 41.7 per thousand and in Rhode Island and Connecticut, where deaths were reported by color, the Negro and mulatto mortality consistently exceeded natality [6, pp. 7-8].

In his classic study, *The Health of Slaves on Southern Plantations,* William Dosite Postell reviews the testimony of contemporary physicians that the Negro slave received "good care, wholesome diet, prompt medical attention, and restraint from dissipations which were injurious to his health" and was, on the whole, "healthier in the main than the whites" [20, p. 143]. Tuberculosis, cancer, scrofula, and syphilis were described as rare among plantation slaves, despite the fact that in the postbellum period the tuberculosis mortality rate was stated to be twice as high among Negroes as among whites [20, p. 143]. Postell's examination of numerous plantation records indicates that slave mortality was about equal to that of the white population in the same areas. The slave infant mortality rate for the plantations studied averaged 152.6 per thousand. By comparison, as late as 1915, the infant mortality rate for Negroes was 163 in Massachusetts, 185 in Pennsylvania, and 192 in New York [21, p. 158].

In conclusion, status, occupation, residence, and freedom for the Negro population of the United States were closely correlated with degree of visible racial admixture. Accordingly, serological indexes of negritude vary from minimum admixture in Southern rural areas characterized by large plantations to maximum admixture in Northern cities, the Negro population of which is partially descended from emancipated slaves and in part selected by the act of migration. On Southern plantations, a selective process operated in favor of those slaves with white genes and those whose appearance and character were approved by their white masters. They were more likely to be employed as house servants or skilled workers, where living conditions were better and mortality was lower. Emancipation was equally selective for Negroes with white genes, but, in this instance, the demographic results were unfavorable. Ignorant of medicine and hygiene, relegated to the lowest, most casual, and worst paid jobs, the free Negroes neither knew how to maintain their health nor were able to afford medical care. Their mortality was consequently much higher than that of plantation slaves, and their rate of natural increase was lower.

Notes

1. The six groups are: southwestern Nigerians, southeastern Nigerians, northern Nigerians, pagan tribes of the Jos plateau in Nigeria, Gold Coast Ewes, and Gold Coast Ashantis.

2. "Narrative, of a five years expedition, against the Revolted Negroes of Surinam, in Guiana, on the Wild Coast of South America; from the year 1772 to 1777: elucidating the History of that Country, and describing its Productions, Viz, Quadrupeds, Birds, Fishes, Reptiles, Trees, Shrubs, Fruits, & Roots; with an account of the Indians of Guiana, & Negroes of Guinea."

References

1. A. E. Mourant. The distribution of human blood groups. Oxford: Blackwell, 1954.
2. Bentley Glass and C. C. Li. Amer. J. Human Genet., **5(1)**:1, 1953.
3. P. L. Workman, B. S. Blumberg, and A. J. Cooper. Amer. J. Human Genet., **15**:429, 1963.
4. William S. Pollitzer, R. M. Menegaz-Bock, Ruggero Ceppellini, and L. C. Dunn. Amer. J. Phys. Anthropol., **22**:393, 1900.
5. W. S. Pollitzer. Amer. J. Phys. Anthropol., **16**:241, 1958.
6. Introduction to 1860 Census of the United States. Washington, D.C.: Government Printing Office, 1800.
7. Arthur R. Jensen. Harvard Educ. Rev., **39(1)**:1, 1969.
8. U.S. Department of Health, Education, and Welfare. Equality of educational opportunity. Washington, D.C.: Government Printing Office, 1966.
9. Sir Charles Lyell. A second visit to the United States, vol. **2.** New York, 1849.
10. Herbert S. Klein. Slavery in the Americas: a comparative study of Virginia and Cuba. Chicago: Univ. Chicago Press, 1967.
11. J. G. F. Burdemann. Notes on Cuba. Boston: Monroe, 1844.
12. Captain John Gabriel Stedman. Narrative of an expedition to Surinam, vol. **2.** 2d ed. London: Johnson, 1813.
13. Bryan Edwards. Civil and commercial history of the British West Indies, vol. **2.** Abridged. London: Bell, 1794.

14. Lewis Cecil Gray. History of agriculture in the southern United States to 1860, vol. 1. Washington, D.C.: Carnegie Inst. Washington, 1933.

15. Joseph Carlyle Sitterson. Sugar country: the cane sugar industry in the South, 1753–1950. Lexington: Univ. Kentucky Press, 1953.

16. Ralph Betts Flanders. Plantation slavery in Georgia. Chapel Hill: Univ. North Carolina Press, 1933.

17. Ulrich B. Phillips. American Negro slavery. New York: Smith, 1918.

18. Louisiana Courier, February 15 and 21, 1840.

19. Kenneth M. Stampp. The peculiar institution. New York: Knopf, 1956.

20. William Dosite Postell. The health of slaves on Southern plantations. Baton Rouge: Univ. Louisiana Press, 1951.

16. Outcrossings in Caucasians and Fetal Loss

JACK B. BRESLER

A rapid increase in outcrossing, or mating between ethnic groups, has been noted in the United States in the past twenty years. There is considerable evidence that there has been a decline in consanguinity rates and a simultaneous increase in the frequency of interethnic marriages. Nonetheless, there have been few studies on the consequences of outcrossing. This report conveys the findings of a research project on the interrelationships among three factors: fetal loss, the number of countries in the background of parents, and the distances between birthplaces of parents.

MATERIAL AND METHODS

Data were collected on every pregnant woman entering the clinic of the Providence Lying-in Hospital, Rhode Island, during the calendar years 1960-1961 and on 131 consecutive pregnant private patients who visited an obstetrician during April and May of 1961.

About 40% of the infants born in Rhode Island and 65% of those born in Providence in the early 1960's were delivered in the Providence Lying-in Hospital. A large number of women residing in the state delivered most or all of their infants at this hospital. Hence, it was possible to obtain a near-complete, and at times complete, reproductive history on many women from the hospital medical records.

However, for local administrative reasons, data on abortions and stillbirths prior to a first live delivery in the Providence Lying-in Hospital, were not available. Therefore, only fertile women, generally in their mid-reproductive stages, who had at least one prior live birth in the hospital, had usable records for purposes of this investigation. This qualification signifies there is no attempt to produce a "standard" for fetal loss in relation to the number of pregnancies. Naturally the women in this study would be older and have had more pregnancies than those in any nationally based set of averages.

From the Providence Lying-in Hospital records, data were collected on parental ethnic groups, place of birth, illegitimate births, twin births, clinical condition, and other factors which could influence the

Reprinted by permission from *Social Biology,* 1970, Volume 17, 17-25.

inclusion or exclusion of the woman and family unit in the sample. Additional and corroborative information on reproductive history of the women was obtained from other hospitals and many gynecologists and obstetricians in Rhode Island.

An in-depth interview was arranged with each woman in the sample to obtain the desired ethnic information. Specifically, each woman was asked to supply information on race, religion, and national origin for: (*a*) herself and her husband (two individuals in the P_1 generation); (*b*) her parents and her husband's parents (four individuals in the P_2 generation); and (*c*) her grandparents and her husband's grandparents (eight individuals in the P_3 generation). Since few women could provide this detailed information immediately, each was asked to complete the survey sheet at home. In addition, each woman was asked to provide additional information about herself and her husband (P_1 generation), such as birthplace, age, prior pregnancies, adoption, and illegitimacy.

Originally, complete records were sought for 792 clinic patients and 131 private patients. Certain items of information were deemed critical, and their absence was sufficient cause for the elimination of the records of a woman and her family from the present analysis. Whole family histories were eliminated if any one of the 42 pieces of the "ethnic" data was unknown. Additional reasons for exclusion from the main sample were: a known illegitimacy in any prior live birth, an incomprehensible reproductive history, a reproductive history involving two or more husbands, and the presence of twins in any pregnancy. Excluded from this report are the 158 family histories where one or more individuals had an African background because interracial matings are not the focus of the present study.

In summary, all matings accepted for this ethnic study met the following conditions: (*a*) all 14 individuals in the family history were white; (*b*) national origin and religion was known for all 14 individuals in the family unit; (*c*) according to the best available information at the time, the pregnant women had never had an illegitimate child, a previous husband, or twins; (*d*) specific birthplaces were known for the pregnant women and their husbands; (*e*) all the pregnant women had at least one prior live-born child. Almost all of the pregnant women and their husbands lived in an urban-suburban environment in central Rhode Island.

The data on reproductive wastage were originally collected in terms of stillbirths and abortions. At that time hospital officials designated reproductive wastage of under 28 weeks as an abortion, and more than 28 weeks as a stillbirth. However, continued experience in handling the data revealed many inaccuracies in assignments compelling us to abandon the classification and substitute the collective term "fetal loss."

The sample presented in this paper contains 708 family histories. Of these, 593 were from clinic patients and 115 were from private patients. These two subsamples were combined after an analysis of two variables: the average numbers of pregnancies and of live births in each group. For clinic patients, the averages were 3.31 and 2.99 respectively; for private patients, the averages were 2.42 and 2.16 respectively. However, the critical comparison is the percentage of live births compared to total pregnancies. The population of clinic patients was unexpectedly higher—90.2% as contrasted with 89.3% for the private patients. A chi-square test was made on private vs. clinic and live births vs. fetal loss. The chi-square value was 0.11 after a correction was made for continuity, with $P > 0.50$.

The second consideration was a comparison of the number of countries of birth in the parental generation (P_3) of the 708 couples. Up to six different countries were found in the parental backgrounds of the 593 clinic couples. By number of countries and number of couples, the figures were: one, 189; two, 228; three, 137; four, 31; five, 7, and six, 1. The numbers for the 115 private patients were: one, 35; two, 45; three, 24; four, 9; five,2; and six, 0. A t test revealed $P > 0.50$.

It may also be argued that the relationship of socioeconomic status to fetal loss represents a bar to combining the clinic and private samples. Table 1 shows the assignment of the families into a tripartite socioeconomic arrangement.[1]

Table 1. Relation of socio-economic status to fetal loss

Socio-economic status index*	Total families	Av. no. pregnancies	% Fetal loss/ pregnancies	% Private patients
Upper	22	2.9	11.1	90.9
Middle	113	2.8	10.0	42.6
Lower	573	3.3	9.8	8.4
Total	708	3.2	9.9	16.2

*Upper combines Index value 1, 2, and 3; Middle, 4, 5, and 6; and Lower, 7, 8, and 9.

It is evident from the sample data that socioeconomic status appears to have little effect on fetal loss (chi-square test: live born vs. fetal loss value is 0.134, 2 d.f., $P > 0.90$). For this and other reasons cited above, it was judged unnecessary to keep the clinic and private samples separated for analysis.

RESULTS

As expected, an analysis of the data clearly shows that fetal loss increases with age. However, in all of the subsequent analyses presented in the tables of this report this factor did not show significant deviations from the mean in any of the classes or groups of data so as to represent a cause for increased or decreased fetal loss.

Perhaps the prime relation is between fetal loss at the P_1 level (signifying the generation at which fetal loss takes place) and the number of countries of birth at the P_3 generation. The data in Table 2 show that with increasing number of countries in the background, there is an ever-increasing level of fetal loss at the F_1 level. When the data for four, five, and six countries were combined in a test involving live born vs. fetal loss, the chi-square equaled 8.87 with the level of probability between 5% and 2%.

Family records were also separated into two categories: (*a*) fetal loss never recorded in the F_1 level; and (*b*) at least one fetal loss in the F_1 reproductive history. The mean number of countries in the "Never" category was 2.024, while in the "Ever" category it was 2.222. A *t* test on the difference of the means shows $P > 0.025$. Accordingly, the two tests on data in Table 2 indicate that more countries of national origin in the P_3 background are associated with greater fetal loss in the F_1 generation.

In Table 3, fetal loss and the number of countries are reviewed again, but this time in terms of maternal and paternal backgrounds. The conglomerate super-cell of information brings together six single cells each having fewer than ten family histories. The 1/1 cell in this table contains more families than are found in the one-country-of-birth category of Table 2. Examination of the data reveals that the 1/1 cell contains many families which would appear in the two-country category of Table 2.

The distance between birthplaces within the continental 48 states of the P_1 generation was charted on a straight-line basis. In this analysis

Table 2. Relation of number of countries in P_3 generation to fetal loss in the F_1 generation

No. countries	No. families	Pregnancies	Live born	% Fetal loss	Never fetal loss	% Families ever F.L.
1	224	736	680	7.6	180	19.6
2	273	901	807	10.3	213	22.0
3	161	471	419	11.0	125	22.4
4	40	114	99	13.2	28	30.0
5	9	28	23	17.9	4	55.6
6	1	2	1	50.0	0	100.0
Total	708	2,252	2,029	9.9	550	22.3

only 646 families were used, since 62 were eliminated because one or both of the P_1 generation members were foreign born. The greatest distance between two points in Rhode Island is approximately 50 miles, and thus

Table 3. Relation of Number of Countries of Birth in Maternal and Paternal P_3 generation and fetal loss in F_1 generation

Maternal no. countries		Paternal no. countries				
		1	2	3	4	Totals
1 ...	No. families	361	81	11		454
	No. pregnancies	1,225	206	32		1,465
	% fetal loss	8.8	10.7	12.5		9.2
2 ...	No. families	146	58			211
	No. pregnancies	449	176			643
	% fetal loss	9.1	15.3	Sum		10.7
3 ...	No. families	25	26			39
	No. pregnancies	78	86			129
	% fetal loss	11.5	14.0			12.4
4 ...	No. families					4
	No. pregnancies					15
	% fetal loss					20.0
Total	No. families	535	149	23	1	708
	No. pregnancies	1,760	414	76	2	2,252
	% fetal loss	9.0	12.6	14.5	50.0	9.9

couples born within the state were placed in a separate category. However, many couples then residing in Rhode Island had birthplaces in other states which were also less than 50 miles apart. These too were put in a cognate category below the "Rhode Island × Rhode Island" (R.I. × R.I.) and indicated as "Elsewhere × Elsewhere" (E. × E.).

In Table 4, increasing distance between P_1 birthplaces is associated with an increase in the average number of countries of birth two generations prior, a time span of approximately 70 years. Data from the two rows indicated as 0-50 were compared in a chi-square test, which showed there was no difference in the average number of P_3 countries between these two populations. For a further chi-square test, data from the last two rows were combined to make one row of 1,000 miles or more. In addition, data in the columns of 4, 5, and 6 countries were combined, making 12 cells of data. The twelfth cell has an expected value below five. Omitting this cell results in a chi-square value of 11.14, somewhat short of the 5% level (6 d.f.); including the cell results in a chi-square value of 16.02, and this is > 1% level.

Table 4. Relation of number of countries in P_3 generation and distance between birthplaces in P_1 generation

Distance in miles	No. countries						Av. no. countries
	1	2	3	4	5	6	
0–50:							
R.I. x R.I.	132	163	104	23	4	0	2.07
E. x E.	9	11	5	2	0	0	2.00
50–999:							
R.I. x E., E. x E.	48	69	35	11	4	1	2.15
1,000–1,999:							
R.I. x E., E. x E.	3	5	8	2	0	0	2.50
2,000+:							
R.I. x E., E. x E.	1	3	2	1	0	0	3.43
Total	192	249	155	40	9	1	2.11

Data in Table 5 show that with increasing distances there is greater fetal loss. Comparing all matings involving less than 50 mile distance with all matings of more than 50 mile distance, and also comparing them on a fetal loss vs. live born basis, the chi-square value was 8.21 or > 1%

after a correction for continuity was made. Admittedly, the relation between distance and fetal loss alone is superficial, but the relation assumes meaning when viewed from the findings indicated in Table 4. The fundamental relation is between fetal loss and number of countries of birth with distance between birthplaces "representing" this more significant association.

Table 6 contains the combined data of Tables 4 and 5. The relationship between increasing fetal loss and countries of birth is corroborated. However, a difference between the two distance groupings is noted. An immediate interpretation cannot be offered for this "migration factor." A subsequent fuller report containing further analysis involving interfaith and intrafaith religious matings, blood group genes, height of parents, and age of parents is planned.

Table 5. Relation of distance between P_1 birth places and fetal loss in F_1 generation

Distance in miles	No. families	No. pregnancies	Live births	% Fetal loss
0–50:				
R.I. x R.I.	426	1,353	1,237	8.6
E. x E.	27	92	81	12.0
50–999:				
R.I. x E., E. x E.	168	526	459	12.7
1,000–1,999:				
R.I. x E., E. x E.	18	39	34	12.8
2,000 + :				
R.I. x E., D. x E.	7	15	11	26.7
Total	646	2,025	1,822	10.0

COMPARATIVE DATA

An earlier report (Bresler, 1961*a*) contained data on blood group genes and infant hemoglobin. Data on reproductive wastage were collected but never published. Those data are now useful for presenting additional evidence on a second population. Some details concerning the population

Table 6. Relation of fetal loss percentages in F_1 generation to distance between P_1 birthplaces and number of countries of birth in the P_3 generation

Countries of birth	Less than 50 miles	More than 50 miles
1	6.7%	9.1%
2	7.9	15.4
3	11.1	13.2
4–5–6	12.8	18.3

have immediate usefulness in this report, and it would be well to summarize them.

Reproductive histories were collected on all white, Rh negative women who entered Providence Lying-in Hospital between the years of 1950 and 1958. No cases were rejected for any clinical disorder. Certain indices of the original data indicated that the population was typical for the geographical area. These were the ABO distribution of mothers, fathers, and infants, and the Rh distributions for fathers. As for the present sample, which we shall designate as Population A, data were collected on women who had at least one prior live birth in the hospital. In the latter part of 1961 we contacted the parents by mail for the ethnic information and received 391 useful histories.

In summary, the following information was available on Population B: (*a*) all mothers were white and Rh negative; (*b*) all fathers were white; (*c*) all members in the P_3 and P_2 generation were listed as white; (*d*) birthplace information was available on all parents; (*e*) a reproductive history was used only when no known or suspected illegitimacy, twins, or two husband situations were indicated; and (*f*) all women had at least one prior live birth.

In Tables 7 and 8 the number of countries of birth in the P_3 generation is related to factors previously analyzed for Population A. Although there are insufficient numbers, it is clear that there are replications of the directions reported in Population A.

In Table 2 fetal loss begins at 7.6% for the one-country category, whereas in Table 7 it begins at 8.6%. Ignoring all country categories where five or fewer families are found, the fetal-loss percentage increases 10.3% in four increments in Table 2 and 8.8% in three increments in Table 7. In Table 2 the approximate average increment is 2.6%, whereas

Table 7. Relation between number of countries of birth in P_3 generation of population B and (a) fetal loss in the F_1 generation and (b) distance between birthplaces of P_1 generation

Variables	No. countries 1	2	3	4	5	6	Total
No. families	123	136	96	30	5	1	391
No. pregnancies	370	428	286	86	18	4	1,192
Live born	338	385	255	71	13	2	1,064
% Fetal loss	8.6	10.0	10.8	17.4	27.8	50.0	10.7

in Table 7 it is 2.9%. Combining these data, the sum of the seven increments is 19.1%, or an approximate 2.7% increase for each country. Hence, it would tentatively appear that each additional country adds between 2.5% and 3.0% fetal loss. At least this appears to be the case in the lower half of the theoretical range of countries from 1 through 8.

TWINS

Twin data for 16 families in Population A and 6 families in Population B were available. The mothers in Population A had 72 pregnancies with an average of 4.5 each. Five pregnancies failed to result in a live birth, resulting in a 6.9% fetal loss. In Population A, the P_1 generation had

Table 8. Relation between number of countries of birth in P_3 generation of population B and distance between birthplaces of P_1 generation

Distance in miles	No. countries 1	2	3	4	5	6	Av. no. countries
0–50:							
R.I. x R.I.	89	79	63	15	3	1	2.07
E. x E.	10	16	8	1	0	0	2.00
50–999:							
R.I. x E., E. x E.	22	32	19	3	1	0	2.08
1,000–1,999:							
R.I. x E., E. x E.	0	4	1	0	0	0	2.2
2,000 + :							
R.O. + E., E. x E.	0	1	1	0	0	0	2.5

a 2.44 average number of countries in the P_3 generation. For Population B the results were respectively: 19 pregnancies, a 3.17 pregnancy average, 5.3% fetal loss, and 2.5 as the average number of countries. If, for the sake of a broad overview, the results from both populations are combined, the results are: 91 pregnancies, a 4.13 pregnancy average, 6.6% fetal loss, and 2.45 as the average number of countries (P_3).

The first observation to be made in comparing these results to those in Tables 2 and 7 is that the average number of pregnancies is higher. Even more unexpected is the markedly low fetal loss of 6.6% vs. 9.9% (Population A) and 10.7% (Population B). Thus, the fetal loss percentage of the twin population is below the "one-country" category for either Population A or B, and there is a higher average number of countries in the twin population, 2.45 vs. 2.07 (A) and 2.13 (B). For the average number of countries, an approximate 10.6% fetal loss is "expected" rather than the observed 6.6%. No explanation can be offered on this twin data combination of high fertility, low fetal loss, and high average number of countries in the background. It would appear, however, that women giving birth to twins have a favorable reproductive history.

DISCUSSION

It is concluded that greater fetal loss is indeed related to greater heterogeneity in the background. This would appear to be the case for two all-white samples in this investigation. No extrapolation of these findings can be made to interracial matings at this time.

The adverse effect of additional countries on fetal loss appears to be cumulative, there being no threshold, in terms of number of countries, which is associated with a large increase in fetal loss. Additional data involving six or more countries in the background are needed to see whether a threshold exists at these levels. The salient question remaining is how to relate this increased fetal loss to genetic and/or social factors.

Previously (Bresler 1961*b*, 1962) I reported on the relationship of the fertility patterns of a white Rhode Island population of one generation and the ethnic mixtures in previous generations. It was shown that the greater the number of national origins in the P_1 background, the lower was the completed mean family size of the F_1 generation. The findings of that study are complementary with those of the two mid-reproductive populations reported here, both of which show increased fetal loss with outcrossings. These collective results can best be interpreted from a genetic basis, with sociocultural factors entering into the findings in a minor way. The most suitable explanation is as follows: The greater the number of countries in the background, the greater is the likelihood that larger numbers of Mendelian gene pools are brought together. With greater mixing of gene pools and concomitant genetic imbalance between loci, fetal loss increases. Fortunately there is an animal model which supports this view as shown by the summary of the excellent series of experiments conducted by Moore (1943, 1946, 1947, 1949, 1950, 1951, 1952) which follows.

In many ways, the evolutionary status of the North American frog, *Rana pipiens,* is similar to the human species. The geographic range of the frog is very wide, encompassing virtually all of North America. *Rana pipiens* is found in many different environments, such as both hot and cold areas, wet and dry areas, and at high and low altitudes. While the taxonomic status is not clear, it seems reasonable to conclude that the species complex contains a very large collection of local geographical races even though it is convenient to regard the whole group as a "single" species. For example, a frequently used taxonomic set of characteristics such as pigmentation shows such a bewildering array of clinical shifts that it is difficult to sort the group into cohesive units (Bresler, 1963, 1964).

Accordingly, *Rana pipiens,* like man, is a vertebrate found in a very extensive geographical range—sometimes isolated, sometimes partially isolated, and sometimes in broad panmictic populations. It is also difficult to classify even with characteristics commonly accepted by investigators.

Moore found that he could bring frogs from different environments of the geographical range into the laboratory, make appropriate genetic crosses, and study reproductive wastage and development. Significant findings include:

1. The hybrids between members of adjacent geographical territories tended to be normal in development and morphology.
2. The greater the geographical distance between parental combinations in eastern North America, the more retarded was the rate of development, the greater were the morphological defects in the hybrids, and the fewer were the normal individuals.
3. The greater the geographical distance between parental combinations, the larger was the percentage of eggs which failed to develop properly.

4. The further apart in geographical distance in eastern North America the members were collected from, the earlier in development did reproductive wastage occur.

Strict comparisons between this study and the previously published excellent study on interracial crosses in Hawaii (Morton, Chung and Mi, 1967) are difficult to make. That study was addressed to first-generation racial hybrids and a complex of indices of heterosis and recombination as well as issues involving quality of live offspring. The present study is an intraracial study of matings involving significantly smaller evolutionary differences between parents and their effect only on one index, fetal loss.

SUMMARY

Data on two white populations show that fetal loss (F_1 generation) in matings of the parental generation (P_1) increases cumulatively by approximately 2.5% to 3% with each additional country of birth in the great-grandparental generation (P_3). A dependent relation shows that increased fetal loss is also related to greater distances between birthplaces of mates within the P_1 generation. Conversely, low fetal loss is encountered with a small number of countries in the background and shorter distance between birthplaces.

It is suggested that a larger number of countries of birth represents a larger number of Mendelian gene pools and that with increased mixture of these gene pools, fetal loss increases proportionately. An animal model is cited in support of this contention.

Notes

1. Each family (P_1 generation) was analyzed independently by Miss Diana Skirmuntas, graduate student in sociology, working under the guidance of Dr. Louis Orzack. The evaluation was based on methodology cited in Hodge, Siegel, and Rossi, 1964.

Acknowledgments

I have been assured by colleagues in the fields of the genetics and physiology of human reproduction that the data provided here are still useful despite the fact that several years have passed since their collection. I gratefully express appreciation for this encouragement and the careful manuscript review by: Dr. V. Elving Anderson, Dwight Institute, University of Minnesota, Minneapolis, Minnesota; Dr. Gordon Allen, National Institute of Mental Health, Bethesda, Maryland; Dr. Melvin M. Ketchel, Department of Physiology, Tufts University School of Medicine, Boston, Massachusetts; and Dr. Jacob Feldman, Department of Bio-Statistics, Harvard School of Public Health, Boston, Massachusetts.

I also thank Dr. Charles Read, obstetrician in private practice in Providence, Rhode Island, for the opportunity to canvass his patients; Miss Dolores Burrow, for the hospital data collection; Mrs. Ruth Tennant, for conducting the ethnic survey for the clinic patients; Mrs. Susan Griffith and Mrs. Veronica Cafferty, for the ethnic survey of the private patients; Dr. Alice L. Palubuiskas, for statistical assistance; Mrs. Gaelen Phyfe, for editorial assistance; and Mr. Mark I. Bresler, for data processing assistance. This study was supported in part by Public Health Service Grant RO1 HD-00603 and by research support from the Graduate School of Boston University.

References

Bresler, Jack B., 1961*a.* The relation of population fertility levels to ethnic group backgrounds. Eugen. Quart. **8**:12-22.

______. 1961*b.* Effect of ABO-Rh interaction on infant hemoglobin. Hum. Biol. **33:**11-24.

______. 1962. The relationship between the fertility patterns of the F_1 generation and the number of countries of birth represented in the P_1 generation. Amer. J. Phys. Anthropol. **20:**509-513.

______. 1963. Pigmentation characteristics of *Rana pipiens:* Dorsal region. Amer. Midland Natur. **70:**197-207.

______. 1964. Pigmentation characteristics of *Rana pipiens:* Tympanum spot, line on upper jaw, and spots on upper eyelids. Amer. Midland Natur. **72:**382-389.

Hodge, Robert, Paul M. Siegel, and Peter H. Rossi. 1964. Occupational prestige in the United States 1925-63. Amer. J. Sociol. **7:**286-302.

Moore, John A. 1943. Study of embryonic temperature tolerance and rate of development in *Rana pipiens* from different latitudes, p. 175-178. *In* Yearbook of the American Philosophical Society, 1942. Philadelphia.

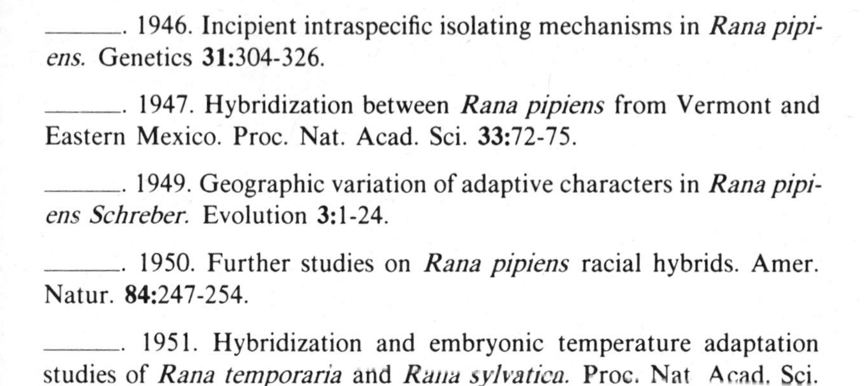

———. 1946. Incipient intraspecific isolating mechanisms in *Rana pipiens.* Genetics **31:**304-326.

———. 1947. Hybridization between *Rana pipiens* from Vermont and Eastern Mexico. Proc. Nat. Acad. Sci. **33:**72-75.

———. 1949. Geographic variation of adaptive characters in *Rana pipiens Schreber.* Evolution **3:**1-24.

———. 1950. Further studies on *Rana pipiens* racial hybrids. Amer. Natur. **84:**247-254.

———. 1951. Hybridization and embryonic temperature adaptation studies of *Rana temporaria* and *Rana sylvatica.* Proc. Nat Acad. Sci. **37:**862-866.

——— 1952. An analytical study of the geographic distribution of *Rana septentrionalis.* Amer. Natur **86:**5-22.

Morton, Newton, Chin S. Chung, and M. Mi. 1967. Genetics of interracial crosses in Hawaii. S. Karger, Basel.

G. Medicine and Law Consider Human Genetics

Human genetic information and methodologies are making considerable impact upon medicine and law.

The twin eugenic themes of a) encouraging the production of offspring from suitable matings and b) discouraging the production of offspring from unsuitable matings occur prominently in the two enclosed papers, the papers suggested in Additional Readings, and in many publications where medicine and law consider human genetics.

In the paper by Lynch and colleagues, genetic counseling is seen as an integral part of medical care and is not distinct from diagnosis, therapy, and prevention of human ills. They encourage the family physician to convey counseling information. This is a most desirable goal but it is regrettable that few American medical schools give genetic counseling prominence.

A legal patchwork of laws has developed in the United States regarding the prohibitions of some forms of inbreeding. In a fascinating review, Farrell and Juberg point out that these prohibitions are based upon different systems of laws and different interpretations of genetic relatedness.

Additional Readings

Wright, Stanley W. and Robert S. Sparkes, 1968. Genetic counseling in mental retardation. *Pediatric Clinics of North America.* **15:**905-923. A good summary paper dealing with mental retardation and its related biochemical, cytological factors in counseling.

Murray, Robert F., Jr., 1969. Genetic counseling in pediatric practice. *Journal of the National Medical Association.* **61:**54-59. Some thoughtful comments on counseling that a pediatrician should consider.

Freire-Maia, Newton, 1970. Empirical risks in genetic counseling. *Social Biology,* **17**:207-212. Cautions on genetic counseling based upon empirical risk information involving averages of different risks operating in different families.

Kindregan, Charles P., 1969. Abortion, the law, and defective children: a legal-medical study. *Suffolk University Law Review.* **3**:225-276. A fascinating legal review dealing with abortions. The author "formulated an opinion against legalization of eugenic abortion (but admits) others may infer different conclusions based on the same data . . ."

Questions

The authors of the genetic counseling paper included in this volume and the authors of the papers suggested as Additional Readings have different opinions on the role of the physician in genetic counseling. How much genetic counseling do you believe a practicing physician should undertake?

If genetic counseling were a requirement for a marriage license, what social effects do you think would occur?

Assume that you are a United States Senator and that you are now considering the revision of laws prohibiting certain marriages: (a) Would you decline to do so? For what reason? (b) Assuming that you do believe there should be such national laws, how could you proceed?

17. Genetic Counseling and the Physician

HENRY T. LYNCH, GABRIEL M. MULCAHY, AND ANNE J. KRUSH

Genetic counseling should be an integral part of medical care when hereditary factors are either suspected or actually proven. In 1964, the World Health Organization Expert Committee on Human Genetics[1] recommended increased training of medical personnel in medical genetics, with particular emphasis on genetic counseling. Again in 1968,[2] it was emphasized that "education of the general practitioner in medical genetics is most important. Teaching should be given at both the preclinical and clinical levels of medical education. It is equally important that medical genetics be dealt with in postgraduate education in refresher courses."

Genetic specialty clinics are not the answer since too few exist to meet the pressing demands for counseling. Furthermore, since our approach to genetic counseling entails diagnosis, therapy, and prognosis of the patient and of his relatives at genetic risk, we feel that the continuity of the overall plan will be best assured when the family physician provides this important service.

Our purpose is to present a variety of common genetic counseling problems, handling and disposition of which should naturally fall within the responsibilities of the family physician.

MENDELIAN INHERITANCE

Autosomal dominant. Autosomal dominant inheritance typically affects individuals through two or more generations. Offspring of an affected parent have a 50% risk of inheriting the particular disorder (Fig 1). Relatives are often fully aware of the existence and distribution of the disease in the family. In our experience, however, they will often have missed the biological implications, ie, they may not appreciate the fact that unaffected individuals do not have a risk for transmitting the disease to progeny or conversely that only affected individuals can transmit the disorder. This is obviously an area in which erroneous conceptions about inheritance may give risk to "blame" or "punishment for past transgressions."

Reprinted by permission from the *Journal of the American Medical Association,* 1970, Volume **211,** 647-651.

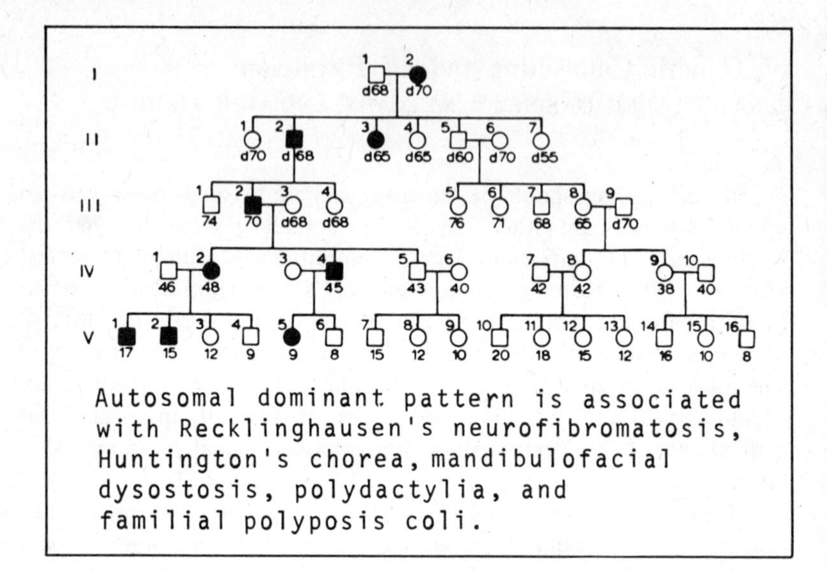

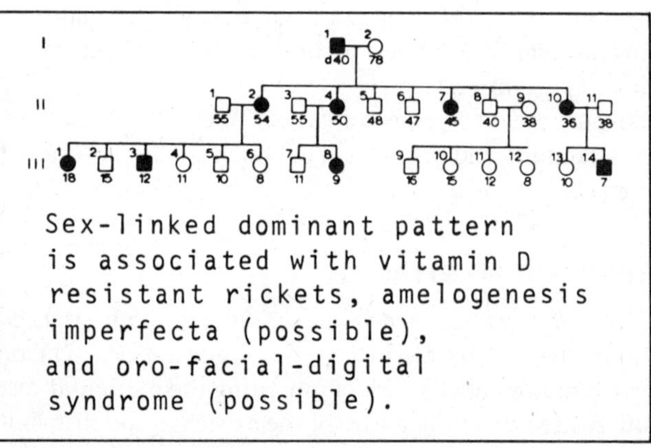

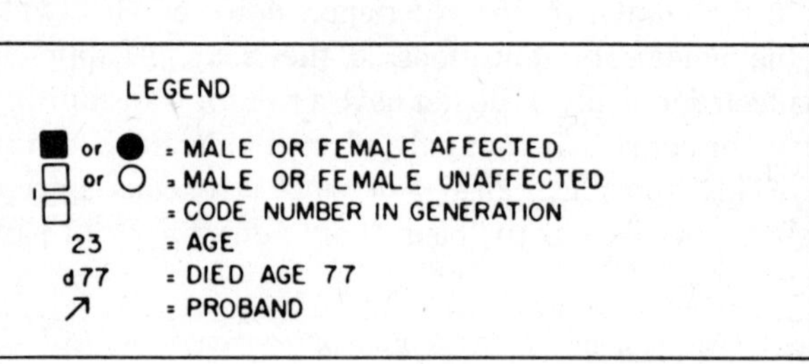

Fig. 1 Hypothetical pedigrees depicting mendelian inheritance

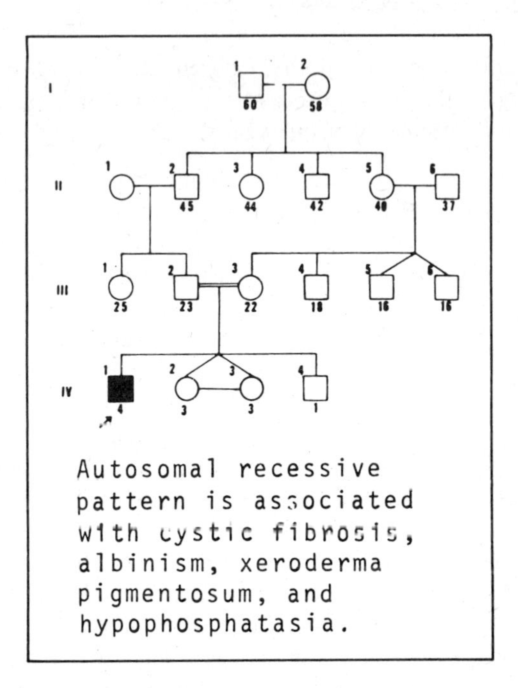

Autosomal recessive pattern is associated with cystic fibrosis, albinism, xeroderma pigmentosum, and hypophosphatasia.

Sex-linked recessive pattern is associated with hemophilia B (Christmas disease), Aldrich syndrome, Duchenne's muscular dystrophy, spastic paraplegia, and Nyhan syndrome.

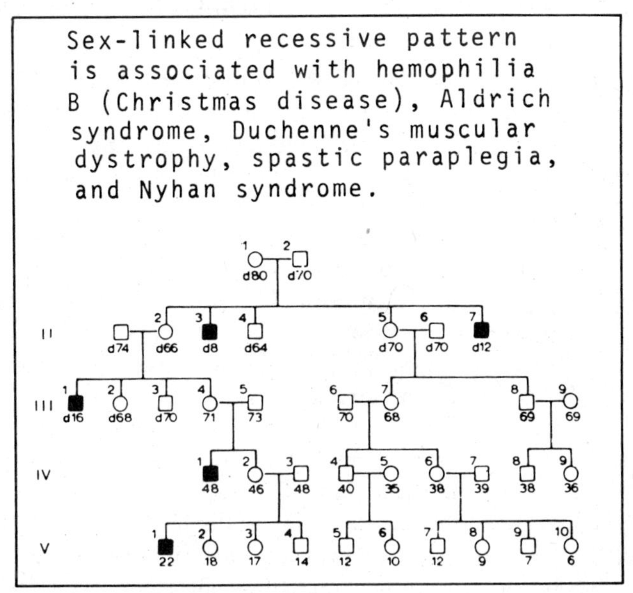

patterns and several examples of clinical disorders in each category.

Community rejection of the kindred often stems from lack of understanding of the mechanisms of inheritance and may hinder full educational and social achievement on the part of family members.[3]

Autosomal recessive. In contrast, the concept of "inheritance" in autosomal recessive diseases may be difficult for relatives to appreciate since affected individuals usually are restricted to a single sibship and their parents are almost invariably normal (Fig 1). Approximately 25% of the offspring of parents, both of whom are heterozygous carriers of the deleterious gene, will be affected. Consanguinity is often present. The biological implications may be profoundly distorted by family members who frequently ask, "How could this have happened to us?" when a child is affected, sometimes followed by self-blame and accusations against the spouse. This type of affective response may be considerably tempered after the parents are informed that they are both heterozygous carriers and that both share in the biological events which led to the disorder in their child. While the risk of only 25% may be considered excessive by some marriage partners, the counselor should emphasize the fact that they have a 75% chance of having a normal child. However, the severity of the disorder, the number of stigmata associated with it, and the future independence and productivity of an offspring also influence the personal evaluation of such risks by the parents.

Sex-linked dominant. Sex-linked dominant diseases occur infrequently in man. The typical case will be that of an affected male who transmits the condition to all of his daughters and none of his sons, or an affected female who has a 50% risk of producing affected sons and daughters (Fig 1).

Sex-linked recessive. In the case of sex-linked recessively inherited diseases, both parents are often normal (Fig 1). Therefore, counseling ramifications may be quite similar to those involved with autosomal recessively inherited diseases. The following is an example of counseling a family with sex-linked hemophilia:

A mother, aged 33, consulted us stating that her maternal uncle had died in his youth from hemophilia and that her sister had two children, both boys, with hemophilia. Our patient exclaimed, "I realize I was taking a chance and was so happy at the birth of each of my two daughters because I dreaded having a son who might be affected like my nephews." Now, however, after practicing contraception for ten years, both she and her husband were eager to have another child and would welcome a boy, if they could be assured that he would not have hemophilia. Her pressing question was, "Can you tell me whether or not I am a carrier of hemophilia?"

To attempt to answer this question, one may make use of both family history and laboratory assay of factor VIII. A certain way of establishing the carrier state is verifying that a mother has already given birth to an affected male child. In this case, of course, there has been no previous male progeny.

Laboratory assay of antihemophilic factor in hemophilia carriers will produce a value in the normal range in most cases, so that a normal value does not exclude the carrier state. In a minority of carriers, however, the concentration will be below the normal range, thus establishing the likelihood that the individual is indeed a carrier. It should be stated that determination of antihemophilic factor is a technically demanding procedure, and only results from a well-supervised coagulation laboratory should be given credence.

It was explained to our patient, therefore, that in her case family history could not answer her question, while laboratory work (on herself and her two daughters) could not exclude the carrier state but might indicate its presence. Coagulation studies on the two nephews would be desired, of course, for verification of diagnosis.

We also informed our patient that without making use of the laboratory tests (which, to date, she has not undergone), her risk for having an affected child is actually one in eight, ie, she has a 50% chance of being a carrier; her chance of having a boy is 50%; and should she have a boy, his chance of being affected is 50%. Therefore, the risk would be $0.5 \times 0.5 \times 0.5 = 0.125$. Of course, there would be an additional one in eight risk of having a female carrier. The patient seemed inclined to accept these risks, probably in part because she had previously believed that the risks of hemophilia were much greater.

CYTOGENETICS

Cytogenetic studies are helpful in disorders in which discernible chromosomal abnormalities of number or structure are present. In most genetic diseases, however, although mutations at the level of the gene no doubt occur, these are far too subtle for direct visual detection with current techniques of chromosomal analysis.

Genetic counseling implications of selected disorders of abnormal or discordant chromosomal constitution

Disorder	Significance to individual	Significance to family
Down's syndrome (21-trisomy)	Mental retardation; associated congenital anomalies; high risk of leukemia	Sporadic cases related to advanced maternal age; familial cases, some related to translocation chromosome or mosaicism in cells of parents; relationship to familial thyroid autoimmunity
13-trisomy (D trisomy)	Death in infancy is expected	Sporadic, related to advanced maternal age; may also result due to translocation chromosome received from phenotypically normal parent
18-trisomy (E trisomy)	Death in infancy is expected	Sporadic, related to advanced maternal age; seldom occurs on basis of translocation chromosome
Cri-du-chat syndrome (partial deletion of short arm of B group chromosome 5)	Mental retardation, small larynx; death may occur in childhood due to associated anomalies or may survive to adulthood	Sporadic, seems not to be related to parental age
Partial deletion of long arm of chromosome 18	Psychomotor retardation; microcephaly; retraction of middle part of face	Sporadic occurrence
Klinefelter's syndrome (XXY, other variants)	Alteration in habitus (eunuchoidism); mentality normal or subnormal; sterility	Generally sporadic; increased maternal age a factor
Turner's syndrome (XO, several variants)	Most XO individuals lost in utero; viable individuals show congenital anomalies, infantile genitalia, short stature, sterility; favorable response to estrogen	Generally sporadic; maternal age seems not to be factor; familial autoimmunity factor
X-polysomy (XXX, other variants)	Most XXX individuals probably normal; some show mental retardation, infertility	Generally sporadic occurrence; offspring of triple X females usually chromosomally normal, but occasional son has Klinefelter's syndrome (XXY)
XYY syndrome	Possible relationship to aggressive criminal behavior; full significance of karyotype uncertain	Generally, sporadic occurrence

Clinical indications for chromosomal analysis include the following: (1) infants and children with mental retardation or congenital anomalies consistent with altered chromosomal number, structure, or suspected chromosomal mosaicism, (2) sterility and infertility problems, (3) differential diagnosis of intersex, (4) psychiatric symptoms with medico-legal implications in males (XYY), (5) problems involving translocation chromosomes and familial occurrences of congenital abnormalities and/or mental retardation, (6) repeated abortions, (7) assessment of chromosomal damage due to virus, x-ray, chemicals, or drugs, (8) hematology problems, as Philadelphia chromosome (Ph^1) in chronic myelogenous leukemia, and (9) rare conditions, such as Fanconi's anemia, in which chromosomal breakage may occur.

The Table summarizes some of the major disorders of abnormal chromosomal constitution with features pertinent to genetic counseling.

In genetic counseling, we use karyotypes as an additional instructional tool for our patients so that they may better comprehend the biological significance of the chromosomal abnormalities. In our experience, this often has a salutary effect on parents. Documentation of the chromosomal change in a severe disease, such as Down's syndrome, often alleviates deep-seated guilt feelings and may end the parents' desperate wandering from one physician to the next in hope of unearthing a mistake in diagnosis.

EMPIRIC RISK PROBLEMS

Empiric risk figures provide an estimate of probability that a particular trait or disorder will occur in an individual under a certain set of conditions.[4] Influences that may determine this probability include genetic and/or nongenetic factors as well as their interaction.

Breast cancer provides a good example of the empiric risk phenomenon. This disease occurs in approximately 5% to 6% of all women. However, a woman, whose first-degree relative had breast cancer, has a threefold increased risk for the development of breast cancer. This has actually been estimated as a 17% risk. Thus, in counseling a patient concerning her risk for breast cancer, one can advise her that hereditary factors play a role, but so far as can be presently determined, a classical mendelian inheritance pattern has not been identified. Empiric risk figures, however, have obvious implications in preventive medicine, since a patient with an increased risk for breast cancer should be taught how to perform a self-examination of the breast and should be encouraged to carry this out monthly at midmenstrual cycle. Breast screening with mammography bilaterally should be carried out annually. In order to avoid cancerophobia in the patient, emphasis should be placed on the cure potential through early diagnosis.

SPECIAL PROBLEMS IN GENETIC COUNSELING

The very nature of some genetic problems may give rise to certain moral, religious, ethical, and legal conflicts. Some of these special problems include adoption (through the means of grey market, black market, or social agency), fertility, infertility, sterility, amniocentesis, therapeutic abortion, medico-legal aspects of teratology, reassurance against occurrence of hereditary defects, etc. With respect to the last named, there are circumstances wherein the counselor is able to *reassure* his patient by providing him with favorable genetic risk information. The following counseling experience depicts the manner in which reassurance was given because of a high level of confidence in the genetic and statistical information at hand.

A family with Huntington's chorea. Genetic counseling was provided to members of a large kindred wherein numerous relatives had succumbed to Huntington's chorea, an autosomal dominantly inherited disorder (Fig 2). A 23-year-old man (Fig 2, IV-1) had recently married. His father (Fig 2, III-5) was 55 and free of any signs of Huntington's chorea. His paternal grandfather (Fig 2, II-9) died at age 77 from a coronary occlusion and so far as could be determined, had never experienced any manifestations of Huntington's chorea. However, ten paternal relatives manifested classical findings of Huntington's chorea (Fig 2). This young man was fearful that this disease might develop; in addition, he and his wife were afraid that their future children would be affected. He was shown that an *affected* individual will have approximately a 50% chance of having *normal* children. He was then shown how normal individuals would beget only normal progeny and that barring a mutation (an exceedingly rare possibility) none of the descendants of a *normal* individual would be expected to have this disease. This individual was reassured that he had no cause for worry so far as Huntington's chorea was concerned since neither his father nor his paternal grandparents demonstrated any evidence of possessing the defective gene. He was greatly relieved and admitted that for the first time in his adult life he could go to sleep without thinking about the dreaded possibility that this disease would develop.

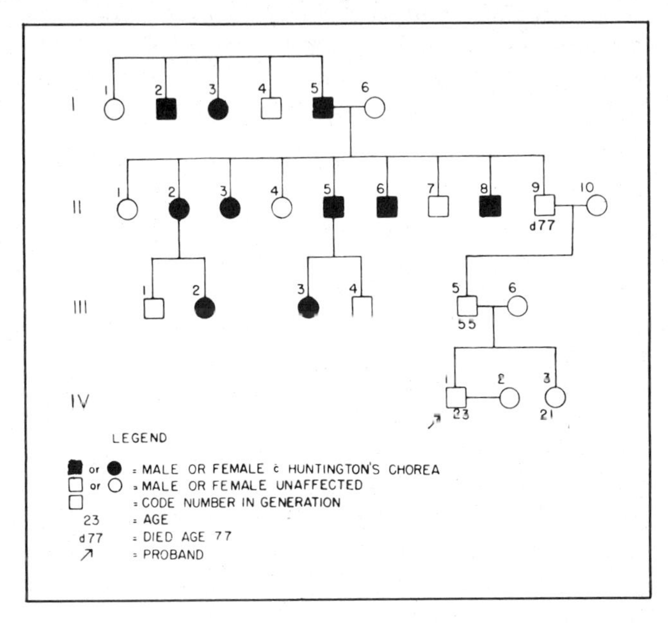

Fig. 2 Pedigree of family with Huntington's chorea.

PREVENTIVE MEDICINE AND GENETIC COUNSELING

In years past, physicians all too frequently have viewed genetic disorders in a fatalistic manner. However, recent advances in medical research, particularly in the area of inborn errors of metabolism, have permitted the control of physical and mental sequelae in some of these disorders. This has been well demonstrated in such diseases as phenylketonuria, galactosemia, Wilson's disease, and others.

Preventive measures have also been extremely effective in other types of diseases including such precancerous disorders as xeroderma pigmentosum[5] and familial polyposis coli.[6] Specifically, restriction of sunlight exposure can be extremely helpful in controlling the serious skin cancer sequelae of xeroderma pigmentosum, an autosomal recessively inherited disorder. In the case of familial polyposis coli, prophylactic colectomy prior to the development of carcinoma can completely alter its prognosis and natural history. The sine qua non for control and prevention in each of these disorders is early diagnosis. This can be readily accomplished through correlating family history with expectations of risk, based on genetic probabilities. It is obvious that stressing preventive aspects of certain hereditary diseases can offer the family hope and often this may facilitate the enlistment of cooperation of relatives who might otherwise attempt to ignore or repress the entire issue.

CONCLUSIONS

Genetic counseling is an integral part of the total management of patients and their families with known or suspected genetic disease. Not only because of an acute shortage of professional genetic counselors, but also because of his knowledge of the medical history of family complexes, the family physician must assume the responsibility for genetic counseling.

In contrast to the frequently quoted comments in the literature which equate genetic counseling with dissemination of genetic risk information, we view genetic counseling as a multifaceted service involving diagnosis, therapy, prognosis, case finding, and prevention, as well as the provision of traditional genetic etiologic and risk information. All of these considerations are of crucial importance in genetic counseling. The family physician is in a position to be in close touch with the family and its problems, and he can often give helpful guidance that takes into account many personal and social factors in addition to purely genetic considerations.

References

1. Human Genetics and Public Health, Second Report of the WHO Expert Committee on Human Genetics, *WHO Techn Rep Ser* **282:** 1-37, 1964.
2. Genetic Counseling: Third Report of the WHO Expert Committee on Human Genetics, *WHO Techn Rep Ser* **416:**5-23, 1969.
3. Lynch, H. T.: *Dynamic Genetic Counseling for Clinicians,* Springfield, Ill: Charles C Thomas, Publisher, 1969.
4. Neel, J. V.: The Meaning of Empiric Risk Figures for Disease or Defect, *Eugen Quart* **5:**41-43, 1958.
5. Lynch, H. T., et al: Cancer, Heredity, and Genetic Counseling: Xeroderma Pigmentosum, *Cancer* **20:**1796-1801, 1967.

6. Lynch, H. T., and Krush, A. J.: Genetic Counseling and Cancer: Implications for Cancer Control, *Southern Med J* **61**:265-269, 1968.

18. Genetics and Laws Prohibiting Marriage in the United States

MICHAEL G. FARROW AND RICHARD C. JUBERG

Laws prohibiting marriage in the 50 states, the District of Columbia, and two territories have been classified as those inclusive for categories of lineal and collateral relatives, and those specific for lineal, collateral, and affinous relatives. A person may not marry a parent, grandparent, child, or grandchild except in Georgia, where a man is not prohibited from marrying his daughter or grandmother. While all political units prohibit marriage between a person and a sibling, an aunt, or an uncle, their prohibitions vary considerably for other degrees of collateral relationship. The uncle-niece marriage is not prohibited in Georgia and among Jews in Rhode Island. Generally, marriage between persons with a coefficient of relatedness equivalent to first cousins or closer has been prohibited. Fewer than one half of the political divisions have prohibitions regarding affinous relatives.

> It iz no crime for brothers and sisters to intermarry, except the fatal consequences to society; for were it generally practised, men would become a race of pigmies. It iz no crime for brothers' and sisters' children to intermarry, and this iz often practised; but such near blood connections often produce imperfect children. The common people hav hence drawn an argument to proov such connections criminal; considering weakness, sickness and deformity in the offspring az judgements upon the parents. Superstition is often awake when reezon is asleep.[1]

So wrote Noah Webster in 1790 in perhaps the first printed discussion on consanguineous marriage in America.

Legal prohibitions of consanguineous marriages in the United States were not declared until the late 19th century and early 20th century when detailed laws were enacted by individual states. Spiritual ideas and practical objectives probably motivated the legislatures more than reliable data and scientific reasoning, although two reports were undoubtedly influential. One of these appeared in 1858 and was presented by Bemiss,[2] who collected data on 873 consanguineous marriages and tabulated the disease and defect in the progeny. The other report was

Reprinted by permission from the *Journal of the American Medical Association,* 1969, Volume 209, No. 4, 534-538.

presented in 1883 by Dr. Alexander Graham Bell[3] who was concerned with the inheritance of deafness. In 1908, Arner[4] authored a monograph on consanguinity in the American population containing a summary of the laws then in effect and analyses of the deleterious effects in the offspring.

The origin of prohibitions of marriage is not actually known. Christian prohibitions to marriage probably date back to Roman Law,[5] although exceptions to the latter law were apparently easily and frequently obtained. If the scriptures formed the only basis for restraints, then at least in some cases, the interpretation was wrong, since Mosaic Law does not prohibit the marriage of cousins. Even further back, Confucius suggested that two people with the same surname should not marry.[6]

The purposes of this report are to tabulate the prohibitions of marriage currently in effect in the 50 states, the District of Columbia, and the territories of Puerto Rico and the Virgin Islands; to illustrate which may be defended by genetic theory; and to show which may be based on legal, moral, or social views.

MATERIALS AND METHODS

We extracted the laws prohibiting marriage from the relevant sections of the most recent edition of the statutes or codes of each of the 50 states, the District of Columbia, and the territories of Puerto Rico and the Virgin Islands. (The specific references may be obtained from the authors.) We also consulted a collection of the laws,[7] although the extractions contain several discrepancies from the original sources. In order to understand the prohibitions in effect in Georgia and for the Jewish population of Rhode Island, we analyzed the prohibitions contained in Leviticus 18.

We discovered that the legal degree of relationship between two persons may depend upon which system of law is employed for the calculation. There are three different systems of law which may be used. One of these is canon law, which is defined as a body of ecclesiastical jurisprudence. Another, civil law, is synonymous with Roman Law and Roman civil law. It is the rule of action which every particular nation, commonwealth, or city has established peculiarly for itself. The third, common law, as distinguished from the Roman law, the modern civil law, canon law, and other systems, is the body of law which was originated, developed, formulated, and is administered in England, and which

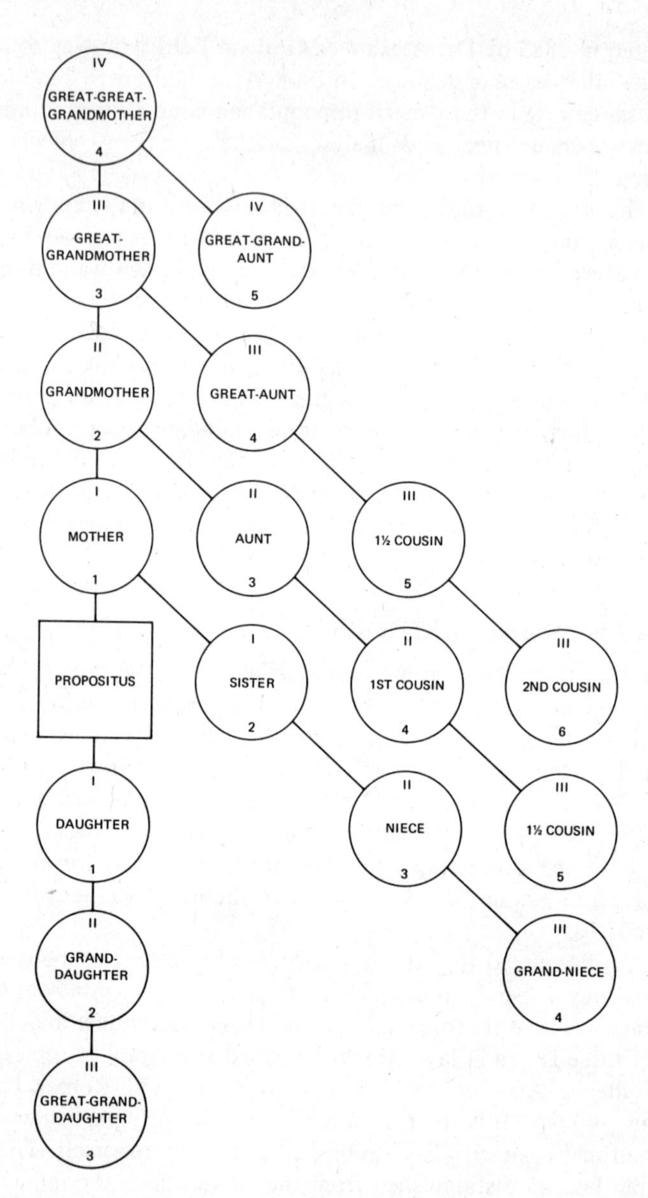

Fig. 1 Calculation of degrees of consanguinity. Arabic numbers refer to degrees computed by civil law, and Roman numerals refer to degrees computed by both canon and common law.

has obtained among most of the states and peoples of Anglo-Saxon stock.[8]

Legally as well as genetically, consanguinity may be lineal or collateral. Lineal consanguinity exists when one person is descended from the other, as between the son and father or grandfather and upward in a direct ascending line, or conversely between the father and son or grandson and downward in a direct descending line. Collateral consanguinity exists when two persons have descended from the same common ancestor but not from each other.

Lineal consanguinity. The degree of lineal consanguinity between two persons is computed by counting the generations in the direct line between them, and the number of links forms the number of degrees of relationship. This method is used by canon, civil, and common law.

Collateral consanguinity. In order to compute the degree of collateral consanguinity between two persons according to canon and common law the common ancestor is first identified. Then the number of generations between the common ancestor and the two persons or the more remote of the two constitutes the degree of relationship. For example, two siblings are related to each other in the first degree, since there is only one step between each of them and the parent. An uncle and his niece are related to each other in the second degree, since there are two steps between the niece and her grandparent, who is the common ancestor (Figure 1). The exact genetic relationship between the two persons is obscured by this method of computation.

By civil law the method of computing the relationship between two collateral relatives is to begin with either and enumerate the steps up to the common ancestor and then down to the other. The total number of steps is equivalent to the degree of relationship. For example, two siblings are related to each other in the second degree. An uncle and his niece are three steps apart and thus are third degree relatives (Figure 1).

Coefficient of relatedness. The coefficient of relatedness may be used to express the degree of genetic relationship between two persons. In general, this may be understood as the proportion of genes which the two persons have in common. The coefficient of relatedness between two collateral relatives may be obtained most easily from the civil law definition. Thus, if there is only one common ancestor, the coefficient of relatedness is ½ raised to the power of the degree of relationship. If there are two common ancestors, then the coefficient of relatedness is 2 X ½ raised to the power of the appropriate degree. For an uncle and his niece the coefficient of relatedness is ¼, while for an uncle and his half niece, it is ⅛. The coefficient of relatedness between two lineal relatives is simply ½ raised to the power of the degree of relationship. For example, a man and his granddaughter are related in the second degree, and the coefficient of relatedness for them is ¼.

While the coefficient of relatedness refers to the probability of genetic similarity between two persons, another term, the coefficient of inbreeding, refers to the probability that at a given locus in one person two genes are identical in their origin. In other words, the coefficient of inbreeding may be understood as the probability that the two genes were derived from the same ancestor. In any event the coefficient of inbreeding (F) for the person may be obtained from the coefficient of relatedness of the parents by multiplying by ½.

Some common terms for relatives do not specify the precise relationship. For example, an aunt may be genetically related to the propositus if she is either his father's or his mother's sister, but she is not genetically related if she is the wife of the brother of either the father or the mother of the propositus. Some states have enacted laws which specifically state the prohibition, eg, in Maine a man may not marry his father's sister or his mother's sister, while he is not prohibited from marrying the wife of his father's brother or the wife of his mother's brother. By contrast, some states simply prohibit marriage of a man and his aunt, eg, Alabama.

A uniform terminology, of course, does not exist among the states. For example, in Pennsylvania a man may not marry his father's wife, while in Vermont he may not marry his stepmother. In the Tables we have combined terms which clearly indicate an identical relationship, in spite of the various terms used in the laws.

RESULTS AND COMMENT

Inclusive laws. We have chosen to use the term inclusive for the laws which establish a general category of relationship within which two people may not marry (Table 1). We have defined two categories of inclusive laws. First, there are laws which prohibit marriage to any ascendant or descendant of any degree. Twenty states have this prohibition, although one state, Maryland, limits the prohibition to within and including these degrees. Thus, in Maryland a man may marry either his great-great-granddaughter or his great-great-grandmother, either of which seem extremely unlikely. The law in Tennessee is more inclusive; it also prohibits marriage to any ascendant or descendant of either parent. The law in Puerto Rico is the most inclusive, since it prohibits marriage to any ascendant or descendant, whether by consanguinity or affinity. Affinity is the relationship established by the marriage of two persons and, therefore, a relative by affinity is not genetically related.

Table 1. Inclusive laws regarding prohibition of consanguineous marriages in the United States and two territories

	A person must not marry			A person must not marry	
State or territory	Any ascendant or descendant	Any relative within and including the following degree	State or territory	Any ascendant or descendant	Any relative within and including the following degree
Alaska		3rd	Nev		5th
Ariz	X		NJ	X	
Calif	X		NM	X	
Colo	X		NY	X	
Conn		3rd	NC		3rd
Del	X		ND	X	
Fla	X		Ohio		5th
Hawaii	X		Okla	X	
Idaho	X		Ore		4th
Ind		5th	SD	X	
Ky		5th	Tenn	X*	
La	X		Utah	X	4th
Md	X†		Wash		5th
Minn		5th	Wis		5th
Mo	X		PR	X‡	3rd
Mont	X		Virgin Islands		3rd
			Total states and territories	20	14

*Or any ascendant or descendant of either parent.
†Within and including three degrees.
‡Or any ascendant or descendant by affinity.

Second, there are inclusive laws which prohibit marriage between two people if they are related within a certain degree; of course, this includes lineal as well as collateral relatives. There are 12 states and two territories with such laws, and two, Utah and Puerto Rico, also have inclusive laws of the first type which prohibit marriage to any ascendant or descendant. Inclusive laws of the second type vary from prohibiting the marriage of any relatives nearer to and including the third degree, eg, an uncle-niece relationship, to prohibiting the marriage of any relatives nearer to and including the fifth degree, eg, a one and one half cousin relationship. Many of the laws are not written the way we have listed them, since they state the closest relationship within which marriage is permitted and then state that marriage between two persons who are more closely related is prohibited. We believe that our listing in Table 1 is easier to understand.

Genetically, the inclusive laws are the easiest to defend, and, legally, they were probably the easiest to write. The genetic reason for advising against the marriage of related persons is, of course, to prevent the coming together in their offspring of any deleterious recessive genes. The probability of this event is determined by the coefficient of inbreeding, eg, 0.125 for an uncle-niece mating, 0.0625 for the mating of first cousins, and 0.0156 in a union of second cousins. Presumably, the role of the legislatures has been to decide what risk is too much to allow the citizens and then to enact a law accordingly.

Prohibitions of marriage to lineal relatives. Contrary to popular assumption and published statement,[7] marriage to first- and second-degree ascendants and descendants is not prohibited in all of the political subdivisions in the United States (Table 2). The exception is Georgia.

Table 2. Prohibitions of marriage to lineal relatives in the United States and two territories

	A man may not marry his				A woman may not marry her			
	Mother	Daughter	Grand-mother	Grand-daughter	Father	Son	Grand-father	Grand-son
Coefficient of inbreeding	1/4	1/4	1/8	1/8	1/8	1/4	1/8	1/8
All states, DC, and two territories except Ga	X	X	X	X	X	X	X	X
Georgia	X			X		X	X	
Total	53	52	52	53	52	53	53	52

The unique law of Georgia lists certain prohibitions of marriage to affinous relatives and then states that marriages in Levitical degrees of

Table 3. Prohibitions of marriage to collateral relatives in the United States and two territories

	A man (woman) may not marry his (her)									
	Sister (Brother)	Half sister (Half brother)	Niece (Nephew)	Half niece (Half nephew)	Aunt (Uncle)	Half aunt (Half uncle)	1st Cousin	1½ Cousin	Half 1st Cousin	2nd Cousin
Coefficient of inbreeding	1/4	1/8	1/8	1/16	1/8	1/16	1/16	1/32	1/32	1/64
Ga	X	X	X*		X*					
DC, Fla, Me, Md, RI, SC, Vt, VI†	X		X*		X*					
Calif, Colo, Hawaii, Mass, NM, NY, Va	X	X	X		X					
Del, Mich, NH	X		X		X		X			
Pa	X	X	X‡		X		X			
Ala	X	X	X	X	X					
Ariz, Ark, Idaho, Ill, Iowa, Kan, La, Miss, Mo, Mont, Neb, WVa, Wyo	X	X	X		X		X			
Alaska, Conn, NJ, PR, Tenn, Tex	X	X	X	X	X	X				
SD	X	X	X		X		X		X	
Okla	X	X	X		X		X			X
NC	X	X	X	X	X	X	X§			
Nev	X	X	X	X	X	X	X	X		
ND, Ore, Utah	X	X	X	X	X	X	X		X	
Ind, Ky, Minn, Ohio, Wash, Wis	X	X	X	X	X	X	X‖	X	X	
Total	53	42	53*	18	53	17	30§‖	7	10	1

*Only aunt-nephew marriage prohibited (Ga) (Jews in RI).
†VI indicates Virgin Islands.
‡Implied by aunt and uncle prohibitions.
§Double first cousins only (NC).
‖Permitted when woman is over 55 yr (Wis).

consanguineous relationship are void. Careful reading of Leviticus 18 reveals that the marriage of a man to either his daughter or to his grandmother is not prohibited; it follows that a woman is not prohibited from marrying either her father or her grandson.

Prohibitions of marriage to collateral relatives. While all political subdivisions have laws prohibiting the marriage of collateral relatives, either specifically designated or else implied through the inclusive laws, the prohibitions vary considerably (Table 3).

The most common prohibitions are those between a person and a sibling, an aunt or an uncle, a niece or a nephew. The prohibitions of marriage between a person and the half relative, ie, the half sibling, the so-called half aunt, half uncle, half niece, or half nephew, all of whom have a coefficient of inbreeding one half that of the full relationship, have not been so comprehensively stated. Thus, among the 53 political units, 42 prohibit marriage between a person and the half sibling, while fewer than half prohibit marriage between a person and the half aunt or half uncle and the half niece or the half nephew.

The marriage of an uncle with his niece is valid for all persons in Georgia and for the Jewish population in Rhode Island. Since the Georgia law on prohibitions of consanguineous marriage depends on the Levitical code, and since this code prohibits the nephew-aunt marriage but not the uncle-niece marriage, then the union of an uncle with his niece is valid. Of course, these two types of marriage are genetically equivalent. The law of Rhode Island is unusual among the laws, since it designates exceptions for a particular religious group, the Jews. The law states that marriages between Jews permitted by their religion are valid. Marriage between Jews in Rhode Island, then is restricted by prohibitions stated in the Bible, the authoritative body of Jewish tradition contained in the Talmud, and in Jewish Biblical tradition.[9] None of these prohibits the marriage of an uncle with his niece.

The intent of the laws regarding the prohibition of marriage between a person and the collateral relative has not been uniformly stated, because popular usage and acceptance of the terms is not the same as the actual blood relationship. Some states have specifically stated that marriage between a person and his aunt by collateral consanguinity is prohibited, or else have stated that marriage between a person and his father's or mother's sister is prohibited. Other states have not been so definite in their intent. At least one state, Oklahoma, specifically declares that the prohibition of marriage to collateral relatives does not apply to relatives by affinity.

The majority of the political units, ie, 30 of 52, prohibit marriage between first cousins, although North Carolina only prohibits marriage between double first cousins and not first cousins. If we consider the political units prohibiting marriage between first cousins plus those prohibiting marriage between persons more closely related than first cousins, we may generalize and state that the risk of 0.0625 of homozygosity at

Table 4. Prohibitions of marriage to affinous relatives in the United States and two territories

	State or Territory							
	Conn, Okla, SD	Ala, Miss, Pa, Tex	Tenn	Iowa NH	Ga	Va, WVa	DC, Me, Md, Mass, Mich, PR, RI, SC, Vt, VI*	Total
A man (woman) may not marry his (her)								
Father's wife (mother's husband)	X	X†	X	X†	X	X	X	23
Grandfather's wife (grandmother's husband)							X	10
Wife's mother (husband's father)				X	X		X	13
Wife's grandmother (husband's grand-father)							X	10
Wife's daughter (husband's son)	X	X	X	X	X	X	X	23
Wife's granddaughter (husband's grand-son)		X	X		X	X	X	18
Wife's stepdaughter (husband's stepson)						X		2
Son's wife (daughter's husband)		X‡	X	X‡	X	X‡	X	20
Grandson's wife (granddaughter's husband)			X	X§			X	13
Nephew's wife (niece's husband)						X‖		2

*VI indicates Virgin Islands
†Father's widow specified (Iowa, NH, Tex).
‡Son's widow specified (Ala, Iowa, NH, Tex, Va).
§Grandson's widow specified.
‖Niece's husband only (Va).

any locus has been thought by the lawmakers to be too much to be undertaken in the population.

Prohibitions of marriage to affinous relatives. Of course, the states which have included restrictions of marriage to affinous relatives in their laws have had legal, moral, or social reasons for the prohibitions rather than genetic arguments (Table 4). The confusion and disagreement which could result regarding the inheritance of property is certainly a legal reason.

Fewer than one half of the states have restrictions of marriage to affinous persons. All prohibit marriage to the spouse of a parent and to the offspring of a spouse. More than one half also prohibit marriage to the spouse of an offspring, to the parent of a spouse, and to the spouse of a grandchild. Some of the other prohibitions are rather curious, eg, the two states which prohibit marriage to a niece's husband and the one state which prohibits marriage to a nephew's wife.

Acknowledgments

Rabbi Herbert J. Wilner, Hillel Foundation of West Virginia University, provided discussion and interpretation of the Jewish laws and traditions. Professor Willard D. Lorensen, West Virginia University College of Law, supplied materials.

References

1. Webster, N.: "Explanation of the Reezons Why Marriage Iz Prohibited Between Natural Relations," *Collection of Essays and Fugitiv Writings on Moral, Historical, Political and Religious Subjects,* Boston: Thomas and Andrews, 1790.
2. Bemiss, S. M.: Report on the Influence of Marriages of Consanguinity Upon Offspring, *Trans AMA* **11:**319-425, 1858.
3. Bell, A. G.: Memoir Upon the Formation of a Deaf Variety of the Human Race, *Memoirs Nat Acad Sci* **2:**177-262, 1883.
4. Arner, G.: *Consanguineous Marriages in the American Population,* New York: Columbia University Press, 1908, pp 13-14.
5. Luckok, H., cited by Arner.[4]
6. Archer, J. C., and Purinton, C. E.: *Faiths Men Live By,* New York: Ronald Press Co., 1958, pp 70-119.

7. Kuchler, F.: "Law of Engagement and Marriage," in *Legal Almanac Series No. 59,* Dobbs Ferry, NY: Oceana Publications, Inc., 1966, pp 21-31.
8. Black, H. C.: *Black's Law Dictionary,* St. Paul: West Publishing Co., 1951, pp 260, 312, 345-346.
9. Adler, C. (ed.): "Incest," in *The Jewish Encyclopedia,* vol 6, New York: Funk & Wagnalls Co., Inc., 1925, pp 571-575.

H.
The Consequences of Social Selection and Genetic Engineering

This section contains three papers; one on customs of child sacrifice of an earlier culture and two on the most modern concepts of genetic engineering. This grouping may seem incongruous. However, all three papers show how man through social customs and scientific techniques has and is altering the genetic constitution of the human population.

Weyl suggests that the custom of the firstborn sacrifice had enormous adverse effects on intellectual pursuits such as literature, philosophy, and science. He notes that few worthwhile materials are available at this time from the Carthaginian culture. Compare Weyl's review on the consequences of the firstborn sacrifice with the statistics on verbal reasoning in the previous reading (No. 9) by Record, McKeown, and Edwards.

The research paper by Behrman and Ackerman was selected as an excellent example of modern man's attempt at genetic engineering. Two terms need to be defined in this paper. Homologous insemination is artificial but with the husband's semen while heterologous is artificial but with semen from someone who is not the woman's husband. The techniques reported upon in this paper have great possibilities in treating infertility.

Huisingh reviews the options given humankind for its genetic future. Throughout the last two papers, many references are made to H. J. Muller. It seems appropriate to begin and end this book with his contributions.

Additional Readings

Davis, Bernard D., 1970. Prospects for genetic intervention in man. *Science,* **170:**1279-1283. A survey paper containing sections on somatic cell alteration, germ cell alteration, cloning, gene transfer, as well as some wise statements on the ramification of genetic engineering.

Markle, Gerald E. and Charles B. Nam, 1971. Sex pre-determination: its impact on fertility. *Social Biology,* **18:**73-83. A survey of students in a university and a junior college indicates that given the choice of the sex of their forthcoming children, an overwhelming majority would choose a boy as the first child and a girl as the second. There is an overall preference of four boys to three girls if sex is indicated for all planned-for children.

Bajema, Carl J., 1971. The genetic implications of population control. *BioScience,* **21:**71-75. He feels that the net effect of current American lifestyles in reproduction is slightly dysgenic in favoring an increase in harmful genes which will genetically handicap a larger proportion of the next generation of Americans. Also, it has a good review of compulsory population program possibilities. Bajema favors a volunteer method of controlling for quantitative and qualitative population growth.

Staffieri, J. Robert, 1970. Birth order and creativity. *Journal of Clinical Psychology,* **26:**65-66. While there is good evidence that firstborn usually have higher intelligence, there are some interesting findings that firstborn are also more conforming, dependent and suggestible than later born. The author feels that firstborn are not as creative as later born.

Petchesky, Rosalind P., 1969. *Issues in Biological Engineering: A Review of Recent Literature.* Institute for the Study of Science and Human Affairs, Columbia University. This booklet reviews recent literature and is an excellent source of information.

Questions

Suppose you could develop population policy, what would you like it to be for 1980? 2000? 5000?

Do you know of other societies which had systematic human sacrifices? What were the probable genetic consequences of it?

Many societies, religious, and ethnic groups forbid certain segments of their population to marry and/or have children. Many such groups on the other hand, encourage greater number of offspring. What are the genetic consequences?

It is inevitable that couples will have the opportunity to choose the sex of their child. Should this be regarded as their right? What do you suppose the social consequences in the United States will be if this is allowed?

Assume that you are a genetic counselor giving advice to a couple. You have determined that they have a high probability of producing a defective child. Despite the overwhelming odds against them, they are strongly inclined to produce a child. Should you as counselor (a) remain neutral, (b) regard the couple as your only responsibility and support their wishes even though you believe them wrong, or (c) regard societal needs as your only responsibility and convey your thoughts to the couple?

19. Some Possible Genetic Implications of Carthaginian Child Sacrifice

NATHANIEL WEYL

When visiting Carthage in October, 1967, I was intrigued by the evidences of infant sacrifice and was tempted to speculate concerning the possible biogenetic implications of this gruesome institution. The evidence that the Carthaginians periodically sacrificed their very young children in vast autos-da-fé to their chief god, Baal Hammon, and later to the goddess Tanit comes from a variety of classical sources and is quite explicit. The most detailed account is that of Diodorus of Sicily, who wrote around 20 B.C., or more than a century after the destruction of Carthage, but who may have relied on earlier sources.

In his novel *Salammbô,* Gustave Flaubert faithfully copied Diodorus's account of a holocaust of neonates in which hundreds of the children of the leading Carthaginian families were incinerated. For this, he was taken to task by contemporary archeologists and historians who asserted that the accounts of this event by Greek and Roman writers should be dismissed as psychological warfare.

Excavations in 1921, however, fully vindicated Diodorus and his popularizer, Flaubert. In the *tophet,* or sanctuary, near the ancient harbor of Carthage where, according to legend, Queen Dido first beached her galleys and later immolated herself on her funeral pyre, crude stelae were discovered. Under these were urns containing the charred bones of thousands of very young children. That this practice was general to Carthaginian civilization and not peculiar to the city of Carthage (*Kart-Hadasht,* or New City) soon became apparent. Thus, since 1963 a cemetery of three thousand sacrificed children from one month to four years old has been under excavation in Sardinia, an area of Punic conquest and settlement. The inscriptions on the stelae and burial urns identify the victims as the first-born sons of noble families and state that they were first strangled and then burned as offerings to Tanit.

Reprinted by permission from *Perspectives in Biology and Medicine,* 1968, Volume 12, 69-78.

NATURE OF THE SACRIFICES

The Carthaginians demanded sacrifice of the first-born of the best families apparently on the theory that human blood was necessary to maintain the supernatural powers of the gods. As a nation of traders, they seem also to have believed that the more valuable the offering, the greater would be the gratitude of their deities. Accordingly, this was not a device for population control or a means of culling the less viable infants, but a practice which must have winnowed out much of the best in the Carthaginian gene pool and operated as a dysgenic factor.

Unfortunately, we lack either statistics or detailed records concerning the extent of child sacrifice, the way the victims were chosen, or the size of the population from which they were drawn. In at least one instance, a deaf and mute child was offered to the gods in return for the gift of a normal child, but sacrifice of the afflicted seems to have been the exception. As to the scope of the practice, we are informed that, when the city was threatened by Agathocles, who invaded Africa in 310 B.C., the priests blamed the calamity on impiety. Many of the leading families had been secretly substituting the children of slaves for their own first-born in the sacrificial holocausts, and there had also been delinquency in payment of tribute to Melkart. Accordingly, five hundred children of the upper classes were put to death in a single auto-da-fé. Thus, the power of the gods was not deemed absolute, for the Punic nobility had dared to deceive them, but it was considered sufficiently great for them to atone by surrendering their own children.

From how large a population were these victims drawn? Strabo estimated that the city had 700,000 inhabitants, but the writers of antiquity were notoriously inaccurate in dealing with large figures. On the basis of the area of the town and the carrying capacity of the surrounding agricultural land, Gilbert Charles-Picard, who headed the excavations at Carthage and directed the Tunisian Department of Antiquities for many years, concluded that the city itself never had more than 100,000 inhabitants and the environs at most another 100,000. If we assume 200,000 for the greater city and make the generous assumption that the "leading families" comprised 5 percent of the total, then the impact of child sacrifice on the demography of the upper classes must have been considerable. Even with such a high birth rate as fifty per thousand, five hundred sacrifices every five years would have exterminated one-fifth of the children of the wealthy. These figures are arbitrary, but they may give some idea of the magnitudes involved.

CONDEMNATION OF PUNIC PRACTICE

Human sacrifice was fairly general in the earlier phases of the Mediterranean civilizations, but was later abandoned everywhere except in the Phoenician cities and in the Carthaginian empire created by Phoenician colonists. Greek and Roman writers consistently condemned the Carthaginians on this score. The Greek practice of exposing unwanted or deformed children did not seem analogous to classic writers. As for the burial alive of two Greeks and two Gauls in the Roman Forum in 216 B.C., this was an almost unique event and a response to the desperate military threat to Rome posed by the defeat at Cannae.

The extent of Greco-Roman abhorrence of Punic child sacrifice is indicated by the fact that Diodorus identified the supreme Carthaginian god, Baal Hammon, not with Zeus but with Cronos, whose chief claim on our memory is that he devoured his own children—hence, the ecological term *kronism* for animal species which control population growth by eating their young.

There are echoes of approval of child sacrifice in the earlier historical books of the Old Testament. In the case of Isaac, the sacrifice is rejected but the offer lauded and considered sufficient reason for Jehovah to make the Jews his chosen people. The blood sacrifice of Jephthah's daughter is accepted by God, but there may be significance in the fact that Jephthah is identified as a Gileadite, not a Jew, and the son of a prostitute.

Solomon made an alliance with the Phoenician city of Tyre and may have reinstituted human sacrifice (I Kings 11:7). Child sacrifice must have become rife in both Judah and Israel after the Assyrian conquest, but it was stamped out by Josiah (*ca.* 620 B.C.), who "defiled Topheth, which is in the valley of the children of Hinnom, that no man might make his son or his daughter to pass through the fire to Molech."

In Carthage itself, the practice may have been somewhat softened with time and increased contact with Greco-Roman civilization. At least the chronologically later layers show charred animal bones mixed with those of infants in the sacrificial urns. Nevertheless, for some reason which is difficult to conjecture, the Carthaginians remained faithful to this bloodthirsty ritual centuries after the complete razing of their city by the Romans, in 146 B.C. Tertullian, a patristic father born in Roman Carthage, wrote indignantly that the surreptitious practice of child sacrifice still continued and that Tiberius, a contemporary of Christ, had tried to stamp it out by lashing the Punic priests to trees and leaving them to die of exposure. Another Christian apologist, Minucius Felix, wrote that the parents who brought their children to the sacrificial altars stifled their children's cries with "kisses and caresses" because the gods did not want weeping victims.

POSSIBLE GENETIC IMPLICATIONS

Two aspects of Punic infant sacrifice may have adversely affected the innate mental ability of the population. The first is that it took its toll primarily from the upper classes. The second is that it affected primarily, if not exclusively, the first-born.

The upper classes in any society are the descendants of those who managed to seize and hold wealth, power, and position. While this may in some instances depend on pure chance, it is more likely to be correlated with mental ability. Furthermore, the upper classes, in human as well as in animal societies, have first choice of females. Thus, selective breeding is continuously operative wherever spouses are selected for brains, character, strength of will, health, and fecundity.

In modern society, the positive correlation between class and innate intelligence is suggested by controlled observations of people of diverse heredity reared together since infancy and of people of identical heredity reared apart. The institutional counterpart of the first is orphanage children, of the second monozygotic twins brought up in different homes.

When he served as psychological consultant to the London County Council, Sir Cyril Burt made a study of orphanage children on the basis of observation and case records. He found to his astonishment that, even when the children had been admitted during the first weeks of infancy and subjected to a largely uniform environment since admission, "individual differences in intelligence, so far from being diminished, varied over an unusually wide range. In the majority of cases, they appeared to be correlated with differences in the intelligence of one or both of the parents" [1]. Among the more striking instances of this rule were orphanage children of high intellectual ability who were revealed by case records to be the illegitimate offspring of fathers of superior social or mental status who had never acknowledged or cared for them. In these cases, superior intelligence could not be attributed to environmental factors. Lawrence (1931) found that orphanage children showed almost as great variability in I.Q. as children of diverse heredity living with their own families, suggesting that the influence of familial environment was secondary. Moreover, the correlation between the I.Q.'s of the orphanage children and the socioeconomic class of their real parents was found to increase steadily with the period of institutionalization [2].

A large and growing literature on monozygotic twins reared apart (and frequently in homes of widely disparate socioeconomic levels) reveals the powerful influence which heredity exerts on intelligence. As summarized by Dobzhansky, a recent review of the data in fifty-two twin studies showed mean I.Q. intrapair correlations of .75 for identical twins reared apart, .53 for fraternal twins, and a mere .23 for unrelated children brought up in the same foster homes or orphanages. The mean I.Q. correlation between foster children and their foster parents was a mere .20 [3, pp. 62-63]. This is illustrative of the comparative magnitudes involved.

Selective slaughter of the progeny of the upper classes could have had just as deleterious an effect on the gene pool as sterilization of the leading families. As for the latter state of affairs, Gilfillan has with great ingenuity traced the decline of Roman invention to the class-selective influence of lead poisoning, which tended to make matrons of the upper classes sterile [4].

We are handicapped in any effort to trace the effects of sacrifice of the first-born of the upper classes on Punic civilization by the dearth of written evidence concerning the latter. We know that the Carthaginians left no architecture worthy of note, that they were grossly inferior in the plastic arts, and were more imitative than inventive even in those fields in which they excelled, such as war, exploration, and trade. They apparently failed to enrich literature, philosophy, or science with anything of consqeuence. At least the only Carthaginian book which has partially survived, and that only in fragments quoted by Greek and Roman writers, is a treatise on agriculture by Mago. If Punic writers had produced anything original or important, there is every reason to believe that Greco-Roman writers would have preserved and quoted from it. In addition, the Carthaginians seem to have been aesthetically underdeveloped and obsessed with death and suicide. Their insensitivity to human suffering was notorious, and they were in the habit of crucifying their unsuccessful generals [5, pp. 71, 98, 203, 204; 6, pp. 58-60].

"The untidiness of the *tophet,* the meagre offerings and the crudity of its funeral monuments," wrote Gilbert and Colette Charles-Picard, "emphasize the aesthetic indifference of the Carthaginians, and their artistic insensibility, and are out of keeping with the atrocious nature of the sacrifices they felt called upon to make. These people who stood so much in awe of God that they suppressed their most natural and human impulses were never capable of giving expression to their religion through the plastic arts" [5, p. 38].

Archeology and history agree in presenting us with the portrait of a somewhat uncouth people, obsessed by *deisidaimonia*, "melancholy and barbaric," odious to their more civilized Greek, Etruscan, and Roman neighbors, neither intellectually creative nor living the life of the mind [6]. This lack of creativity explains both the dearth of historic records and our ignorance of Punic social and intellectual history. We do not know whether superstition, blindness to beauty, and intellectual sterility of the Carthaginians were constant factors in their history or the end result of a process of intellectual decline caused in part by dysgenic child sacrifices which progressively impoverished their genetic heritage.

BIRTH ORDER AND INFANT SACRIFICE

An intriguing question is whether the choice of the first-born for sacrifice was also dysgenic. The fact that the first-born tend to be markedly more successful, more eminent, and more highly concentrated in the academic profession in advanced modern societies has been established by a long line of investigators from Francis Galton through Havelock Ellis [7, p. 103], Ellsworth Huntington [8, p. 292], and Corrado Gini [9], to more recent workers in the field [10; 11, p. 3]. The literature on birth order in relation to achievement, intelligence, and character structure has been ably summarized in a recent article by Altus, who concluded: "Ordinal position at birth has been shown to be related to significant social parameters, though the reasons behind the relations are as yet unknown or at best dimly apprehended" [12].

A question relevant to the dysgenic implications of Carthaginian child sacrifice is whether the differences in intelligence between the first-born and other siblings are wholly due to environmental factors or only partially so. Investigators have stressed such social factors as that the first-born is normally subjected to greater parental strictness than his younger brothers and sisters and enjoys a high degree of parental attention and interaction during the period when he is an only child.

We do not have enough evidence to judge whether these and similar environmental factors operated in favor of the first-born in Carthage. Even if one assumes that these factors did operate with sufficient force to give the first-born a significant advantage in intelligence and achievement, this question would still remain unanswered: Would the first surviving child—that is to say, the oldest sibling of the first-born offered to Baal Hammon—have enjoyed the same preferential parental treatment as his slaughtered sibling?

If the observed differences in psychometric intelligence between first-born and subsequent births are partially due to genetic factors, then the birth order of sacrificial victims must have been relevant to the impoverishment of the Punic gene pool. Evidently, in those instances in which medical complications prevented further pregnancies, sacrifice of the first-born eliminated all the genes from the mating of the victim's parents. My attention has been drawn to a possible second causal factor, namely, that in instances of blood-group incompatibility, particularly Rh, the first-born is more likely than his siblings to escape unscathed and that the probability of maternal isoimmunization and erythroblastosis fetalis is correlated with birth order. A third set of factors is the tendency of chromosomal abnormalities, such as trisomies, to increase with the age of the mother. There is also some evidence that point mutations are more frequent in the sperm of older fathers. Perhaps the pathologically most prevalent trisomy is Down's syndrome, or mongolism, a congenital affliction occurring in more than one-tenth of 1 per cent of all births and causing amentia. It had long been known that mongolism occurs more frequently among the children of older mothers, and it had therefore been argued that the syndrome was due to environmental factors. The discovery that monozygotic twins are concordant for Down's syndrome, whereas dizygotic twins are much less so, forced abandonment of this view. In recent years, geneticists have established that mongoloid children have three chromosomes 21 rather than the normal two. This still does not explain why mongolism occurs with proportionately greater frequency when mothers are older. A suggested explanation is that the female's ability to reject imperfect gametes may decline with age or order of birth [13]. In a letter in *Nature* on the relationship between sperm redundancy and chiasma frequency, Jack Cohen hypothesized that internal fertilization is "primarily an opportunity for sperm selection, so that only 'perfect' gametes are offered at fertilization" [14].

Blood-group incompatibilities and trisomies would tend to produce mentally retarded and grossly defective children. Sacrifice of the first-born would raise the proportion of defectives to total population. Given the predominance of males in the leadership of ancient societies, another contributory adverse factor would be the slight decline in sex ratio with increasing birth order.

The genetic factors mentioned operate primarily by increasing the proportion, and hence the burden, of defectives. They could not be more than a minor contributory cause of the observed superiority of the first-born in achievement and psychometric intelligence. If the main causal factors at work are overwhelmingly social and environmental, one would expect to find first-born superiority more evident in respect to achievement than I.Q. The reason for this is based on the belief that psychometric intelligence reflects innate intelligence more and environment less than does the achieving of eminence or status in life. No tacit assumption is involved that I.Q. tests are culture-free or that they measure genetic intelligence exclusively.

Consider rosters of achievement and eminence first. In 1938, Huntington analyzed those 1,210 Americans, living and dead, whom he considered most worthy of fame. Of those from two-child families, 64.1 per cent were first-born. A study of 235 Rhodes scholars from two-child families revealed that 61.3 per cent were first-born; an analysis of *Who's Who in America* entries produced a corresponding figure of 64 per cent; a study of 1,817 college students from two-child families on the Santa Barbara campus yielded 63 per cent first-born [8, 12, 15]. The unweighted arithmetic mean of these four indexes is 63.1 per cent.

A comparable analysis of the relation of primogeniture to high intelligence is the unpublished paper of R. C. Nichols on 1,618 National Merit finalists [16]. Nichols writes that these selected high school students scored "almost three standard deviations above the mean of the general population." In psychometric intelligence, they probably rank in the first of 0.5 per cent of the American population and are, therefore, a considerably more select group than Terman's California gifted. Of the 568 finalists from two-child families, 66 per cent were first-born. This is more impressive than the average of four indexes cited above. Amazingly enough, the first-born constituted 52 per cent of the finalists from three-child families, 59 per cent of those from four-child families, and 52 per cent from five-child families. Almost 60 per cent of the finalists from families with two to five children inclusive were first-born.

The pre-eminence of the first-born is much less marked when one considers activities requiring merely moderately above-average intelligence. Among 4,300 University of California undergraduates, who must rank in the first 10-15 per cent of their high school classes in grades to qualify for admittance, only a small excess of first-born was noted [12]. The performance superiority of the first-born seems to be markedly higher in the scholastically more exacting colleges than in less exigent institutions. At Reed College, 66 per cent of a sample were first-born; at Yale, 61 per cent; but, at the University of Minnesota, only slightly more than 50 per cent [12].

It thus appears that the advantage of the first-born tends to be larger in psychometric intelligence than in achievement and tends to be greatest at the higher intellectual levels. On the basis of the environmental causal explanations offered, neither of these differences would have been expected.

We do not know what the causes of first-born superiority are, nor do we know the extent to which they are hereditary and the extent to which they are environmental. Hence, it would be illegitimate to assume that the pattern which seemingly emerges from contemporary American studies is applicable to Punic civilization. If the pattern is applicable, however, slaughter of the Carthaginian first-born would not merely have thinned the ranks of the ruling class and to that extent impoverished the gene pool in respect to intelligence, but would have acted, by reason of birth order, to strike down a disproportionately large number of the most gifted progeny of this upper class. This suggests the possibility that the enormous stress placed upon birth order by the Hebrew patriarchs and by many other prescientific societies may have reflected more than merely legal and testamentary considerations. We might find that this was one of those instances in which folk traditions were based on sound empirical inferences from the collective experience of tribe or nation and that the reason for insisting that the first-born inherit was a well-grounded belief that they were likely to be more capable than their younger brothers.

References

1. Cyril Burt. Amer. Psychol., **13** (January):1, 1958.
2. E. M. Lawrence. Brit. J. Psychol. Monogr. Suppl. 16, 1931.
3. Theodosius Dobzhansky. Heredity and the nature of man. New York: Harcourt, Brace & World, 1964.
4. S. C. Gilfillan. J. Appl. Nutrition, **19**(3, 4):95-99, 1967.
5. Gilbert and Colette Charles-Picard. Daily life in Carthage. New York: Macmillan, 1961.

6. Henri Paul-Eydoux. The buried past. 1st ed. New York: Praeger, 1966.
7. Havelock Ellis. A study of British genius. Rev. U.S. ed. Boston: Houghton Mifflin, 1926.
8. Ellsworth Huntington. Season of birth, 1st ed. New York: Wiley, 1938.
9. Corrado Gini. J. Hered., **6**:37, 1951.
10. J. M. Cattell. Sci. Monthly, **5**:371, 1917.
11. A. Roe. Psychol. Monogr. 352, 1953.
12. William D. Altus. Science, **151**:44-49, 1966.
13. Curt Stern. Perspect. Biol. Med., **10**:500-506, 1967.
14. Jack Cohen. Nature, **215**:862, 1967.
15. F. L. Apperly, J. Hered., **30**:493, 1939.
16. R. C. Nichols. Birth order and intelligence (unpublished).

20. Freeze Preservation of Human Sperm

S. J. BEHRMAN AND D. R. ACKERMAN

For one hundred years,[8] all efforts to preserve spermatozoa by freezing to very low or ultralow temperatures have been dominated by a single motive, i.e., simply the desire to control fertility. This motivation is clearly acknowledged with respect to the semen of domestic animals and implied, if not clearly stated, in the efforts of those who have worked with the freeze preservation of human semen.

The particular technique of controlling fertility by means of cryogenic methods has two very powerful advantages: (1) the control of fertility can be established independent of time and place and (2) fertility can be controlled, to a greater or lesser extent, with respect to desired qualities in the offspring. Thus, semen specimens which have been frozen to, and maintained at, ultralow temperatures can be shipped any place and used at any time and may be administered without consideration of the many factors which normally operate during the usual procedures of artificial insemination. It is clear that the rationale of absolute fertility control is well established in the thinking of animal reproduction scientists. It is not clear that this logic of cryogenic technology has been thoroughly thought through by clinicians who are interested in human infertility and in the control of human fertility. The major purpose of this paper will be to present in a systematic way the logic of cryogenic applications to human fertility.

Three major considerations have determined the development of present technology. In the first place, it has been necessary to provide for the total suspension of metabolic activity among treated sperm cells; and it is equally necessary that the indefinite suspension of cellular activity be consistent with the preservation of essentially normal metabolic activity once the cells have been thawed. The second consideration has been the development of this technology in such a way that large amounts of material may be handled with efficiency with respect to dilution, cooling, storage, thawing, transport, and final administration to recipient animals. The third consideration is, of course, the attainment of the highest

Reprinted by permission from *American Journal of Obstetrics and Gynecology,* 1969, volume 103, 654-664.

possible fertility from this material, which is dependent on the satisfactory development of the first two criteria.

At the present time, a freezing preservation technique is applied with enormous success in the control of bovine fertility. Success with respect to other domestic animals has been indifferent or lacking entirely. Again, with respect to the use of frozen preserved semen in human recipients, success has been satisfactory but hardly outstanding. This means, among other things, that the particular requirements which determine developments in animal husbandry are largely self-limiting, and that it is time for human biologists and for clinicians who are interested in human fertility to see the necessity of continued technologic development which is determined by the specific requirements of human physiology and by the ethical and philosophical considerations which apply specifically to the human realm.

For practical purposes, there is a general similarity in cryogenic techniques as applied to human and to animal spermatozoa. But it seems clear to us that there are serious limitations in transferring assumptions and techniques wholesale from animal husbandry to the control of human fertility.

It is absolutely necessary at this point that we realize fully the very large difference existing between species with respect to the ability of their spermatozoa to survive what we think of as standard freeze preservation technique. We have paid lip service to our understanding that there are significant differences between freeze response of bovine and human semen, but the fact that we continue to apply bovine cryogenic techniques simply indicates that we have made no serious effort to discover the nature or the significance of specific difference in this context. Thus, it is mandatory that we begin to formulate questions in terms of the cryogenic requirements of the *human male gamete* rather than in terms of mammalian gametes in general. Briefly, there are three areas of deficiency pertaining to the application of cryogenic techniques to human gametes: (1) How can human semen be frozen and thawed in a way which is consistent with very high fertilizing potential? (2) What will be the biologic consequences to the human species of the application of this technique? (3) What will be the sociologic consequences of the application of cryogenic technique at any given frequency level of application in human population? If these three questions are to be answered, it is clear that we are not going to learn much more of value by continuing to follow the development of this technology in terms of animal husbandry and must concentrate on the human.

In keeping with this tenet, we present the experience of such studies in our laboratory to date.

MATERIAL AND METHODS

These have been described in detail elsewhere,[4] but in brief, the semen samples used for heterologous inseminations were collected from young and healthy students after abstinence for more than 4 days. Samples were also collected once a week or every other week from the husbands of childless couples and stored for inseminations. Most of the latter samples were oligospermic or had a high percentage of immature cells. Semen samples from these husbands were usually obtained by means of split ejaculation, and the first fractions, which usually had a high sperm density, were pooled. Some samples, however, were collected by the conventional single ejaculation and concentrated by centrifugation at 1,500 to 2,000 r.p.m. for 10 minutes before processed for freezing.

After a specimen had been allowed to liquefy at room temperature (20-23° C.) a part to be stored was mixed in a 1:1 ratio with a protective medium, also at room temperature. The protective medium consisted of: (1) egg yolk, 20 per cent by volume; (2) glycerol, 14 per cent by volume; (3) a mixture of 5 per cent glucose aqueous solution and 2.9 per cent sodium citrate aqueous solution in proportion of 2 and 3 volumes, 66 per cent. To each milliliter of the final mixture there were added 1,000 U. of penicillin G, 0.5 mg. of streptomycin, and 20 mg. of glycine. The medium was then heated for 30 minutes at 56° C., adjusted to pH 7.2-7.4 (by smaller volumes of 1.3 per cent $NaHCO_3$ solution than 0.05 of the final mixture and, if necessary, by 0.1N NaOH solution), and stored in a refrigerator. The medium was freshly prepared each week.

After a semen sample had been mixed with the medium the mixture was divided in 1.2 ml. glass ampules so as to provide 1 ml. per ampule, sealed by flame, and then placed in a biologic freezer.* Cooling was initiated not later than one hour after collection of a specimen. The cooling schedule employed in this series was as follows: 1° C. per minute from room temperature to 2° C.; 5 to 7° C. per minute between 2 and –20° C. so as to pass through the range of heat of fusion within 3 minutes;

*Linde Division, Union Carbide Corporation, New York, New York.

and then approximately 10° C. per minute down to −80° C. Immediately after the temperature of the samples reached −80° C., the ampules were transferred to a canister in a BF–25* refrigerator. Temperature of the samples was measured by using a thermocouple placed in the center of the control mixture, consisting of an equal volume of a 1:1 mixture of saline and the medium. Thawing was performed in a 37° C. water bath, until the last ice particle in the ampule had disappeared.

Inseminations were performed within 30 minutes of thawing the samples. One milliliter of the semen sample from a particular donor collected on one occasion—that is, 2 c.c. of a semen-medium mixture—was used for each insemination.

Rapid cooling of human semen above the freezing point damaged respiratory activity of the spermatozoa more severely than their motility. Thus, a rapid rate of cooling at the range of latent heat of fusion seemed to be preferable to protect the spermatozoa from damage at this range. A medium rate of cooling, 5 to 7° C. per minute, was best for protection of the spermatozoa at the range between termination of heat of fusion and −30° C., and slower thawing methods seem to be preferable to faster ones.

RESULTS

As of this presentation only 13 patients have had homologous inseminations and of these two became pregnant. This group has now been discontinued because of poor results. In the heterologous insemination group, Series I constitutes 68 patients, of whom 29 or 42.6 per cent became pregnant, and the rest have now been discontinued from the study. (This compares to a 69 per cent pregnancy rate in 362 cases of donor insemination using fresh semen.)[3] There are 24 additional cases, Series II, under study with frozen donor sperm of which none have yet conceived (Table 1).

A comparison of these data with available literature is presented in Table 2. The number of months it took for these 29 cases to become pregnant is depicted in Fig. 1. Twenty-seven of 29, or 93.1 per cent, were pregnant within 7 months. Of the 29 pregnancies, 28 went to term, including one set of twins. There was one spontaneous abortion and no detectable gross anomalies in the offspring. It took an average of 6.9 inseminations over a period of 4.2 months for these patients to conceive (Table 3).

Table 1. Report of pregnancies obtained in homologous and heterologous inseminations using frozen sperm

	Patients (No.)	Pregnant (No.)
Homologous (A.I.H.)	13	2
Heterologous (A.I.D.)		
Series I	68	29 (42.6%)*
Heterologous (A.I.D.)		
Series II	24	–†

*In comparison, note 68.4 percent success, with fresh sperm.
†Still under study.

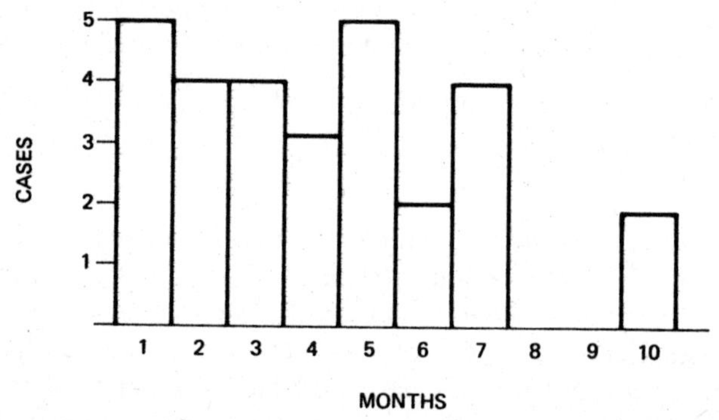

Fig. 1 Number of months necessary for 29 patients to become pregnant with frozen sperm (27 or 93.1 percent in 7 months as compared to 92.8 percent in 6 months with fresh sperm).

Because of previous interest[13] in the significance of prefreeze and postthaw motility of spermatozoa to pregnancy rate, the semen characteristics of specimens known to be responsible for pregnancy were evaluated. The recovery rate of spermatozoal motility was expressed by dividing the prefreeze motility by postthaw motility. From Table 4 it will be noted that the average recovery rate of motility after freezing and thawing was 69 per cent. The average number of days of sperm storage was 49 days (range of one day to 9 months).

Table 2. Summary of the literature on the clinical use of frozen-preserved human semen

Workers and year	Freezing procedures			Results
	Final concentration of glycerol or additives (%)	Cooling speed	Storage temperature (°C.)	
Bunge, Keettel, and Sherman 1954[5]	10	Rapid (medium)	−70	4 births from ? A.I.D.
Keettel, Bunge, Bradbury, and Nelson, 1956[7]	10	Rapid (medium)	−70	9 pregnancies in 26 A.I.D.
Ferandez-Cano, Menkin Garcia, and Rock, 1964[6]	5	Rapid	−79	3 pregnancies in ? A.I.D.
Perloff, Steinberger, and Sherman, 1964[11]	10	Rapid	−196.5	6 pregnancies in ? A.I.D.
Behrman and Sawada, 1966[4]	7.5 (in yolk buffer solution)	Slow	−196.5	12 pregnancies in 28 A.I.D.; 1 pregnancy in 7 A.I.H.
Present series. 1968	7.5	Slow	−196.5	29 pregnancies in 68 A.I.D.; 2 pregnancies in 13 A.I.H.

Table 3. Details and outcome of 29 pregnancies with frozen sperm

No. of months to conceive	4.2
No. of inseminations	6.9
Normal spontaneous deliveries	28 (1 Twin)
Abortions	1
Ectopic pregnancies or anomalies	0

Table 4. Semen characteristics of specimens known responsible for 16 pregnancies (frozen sperm)

Patient Code No.	Sperm recovery* rate (%)	Storage	
		Months	Days
546	86	9	0
551	63		18
593	87		18
0	65		4
611	62	1	0
552	77	1	9
488	93	1	7
533		1	0
461	56		5
579		1	0
579	49	5	11
542	77	2	8
555	53		6
569	81		14
600	54	1	11
568	60		1
Average	69		49

$$\text{*Recovery} = \frac{\text{Prefreeze motility}}{\text{Postthaw motility}} \times 100.$$

COMMENT

From a clinical standpoint it seems quite clear that human spermatozoa can be preserved for considerable periods of time. Conception with this technique requires almost the same number of inseminations as when one uses fresh sperm but the efficacy, or conception rate, is definitely lower when using frozen sperm, i.e., 42.6 versus 69 per cent. However, of those that will conceive, 93.1 per cent do so in 7 months. It is therefore apparent that current techniques are not yet sufficiently sophisticated to obviate some of the changes that are caused by ultralow temperature treatment.

It might be well, therefore, to examine and discuss the future or indeed, the need for extensive investigations into freezing of human spermatozoa.

Systematic banking of frozen human semen is obviously of value in the conduct of a great many essential types of basic investigation.

A number of important studies are dependent upon the availability of human semen which has been preserved for greater or lesser length of time. The in vitro study of cellular aging, efficacy of various storage temperatures, the maintenance of enzyme activity and enzymatic alterations, ice crystallography, biochemical stability or instability, and functional metabolic alterations must all be referred to in the context of time in storage.

It is possible that considerable thought, as well as some technologic advances in semen banking will be necessitated by current concepts and advances in the field of family planning. Clinicians have displayed considerable interest in the idea that large psychological as well as practical advantages would ensue if family planning were conducted on the basis of vasectomy, preceded by the freezing preservation and storage of a sufficient number of the individual's semen specimens to insure that he may have in the future as many children as are desirable. This method of family planning offers a good many advantages. It is clear to those who have worked with semen preservation, however, that our present techniques are neither certain enough, not efficient enough, to allow us to utilize this technique on a widespread basis.

It has been remarked occasionally, in discussions of the practical uses of available frozen human semen, that in the event of an atomic catastrophe of greater or lesser proportion, that it would be valuable to have a significant amount of semen available which had been protected against such an accident. It is true that, in general, semen specimens which had been frozen and banked would suffer considerably less irradiation damage than would the spermatogenic apparatus of the individual males who had been caught within the effective radius of ionizing radiation. It is certainly possible that the stock piling of human semen against some such contingency is at least as valid an enterprise as the stock piling of atomic weaponry, of food, or of any other valued resource. It occurs to us, however, that the availability of a technology for long-term semen preservation has other uses which are of immediate value for both research and clinical purposes.

Work reported from our own laboratories has indicated yet another important research problem, that of donor selection. Actually it is clear that there are many problems in the selection of donors; and only some of these will be discussed. For present purposes, we wish to emphasize particularly the problem of selecting donors whose sperm cells are able to withstand the freeze-thaw process. Our work has demonstrated that we have, in fact, no satisfactory method of prognosticating the fertilizing capacity of semen specimens after they have been treated at ultralow temperatures. We have been able to select some donors for clinical purposes and to discard others as unsatisfactory to these purposes, only on the basis of repeated clinical trials with their specimens. This is grossly inefficient and indicates that we possess at this time no significant understanding of the biologic factors which entail success or failure with respect to the low temperature properties of human spermatozoa. The factors, which must be at least in part genetically determined, which determine "freezability" must be satisfactorily determined before the large scale application of frozen sperm can be successful. We do not doubt ourselves that a solution of many of these factors is entirely within our capabilities on the basis of both in vivo and in vitro work and selected clinical trials; but, determination of new classification procedures for both donors and recipients, combined with the resulting fertility data from clinical trials, will be required.

Many of the topics discussed already in this paper would appear to be, in fact, relevant for future consideration in the development of human seminal cryobiology. This is simply because most of the significant work in this realm has yet to be done. Nevertheless, we have classified these topics as relevant to the present and not to the future, based upon the conviction that problems discussed so far can be attacked and solved at the present time.

Under normal circumstances, when conception takes place, each parent has passed on to the newly fertilized cell one half of its genes. These have been chosen at random, one out of every pair of each parent, and the corresponding genes have again paired up in the newly created cell. The progeny has then received half of its genetic code from each parent, not as a mixture, but in discrete units selected at random, each having its special effect on one or more aspects of development and function. Because of the variety of different types of genes scattered throughout the population, the system of bisexual mating makes for an almost infinite number of possible genetic codes and thus provides the material for change. The potential for change is further increased by mutations,[10] and it can be assumed that there will be further changes effected by low temperature treatment of spermatozoa.

Several very reasonable assumptions are involved here. These assumptions are systematically being explored in our laboratories. They are (1) that the mature ejaculated spermatozoa possess a phenotype which is in part determined by the haploid genetic constitution of the cell; (2) that spermatozoa bearing different haploid genetic constitutions will respond differentially to freezing preservation treatment; and (3) that the variations in freezability between the semen of various donors is, in part, determined genetically. Thus, there is opportunity for *natural selection* as exerted by the pressures of freezing preservation to select among the spermatozoa of a given semen specimen and, in addition, for selection to operate between donors with respect to the net fertility among many recipients of their specimens. For example, Ackerman[2] has produced in vitro evidence that the fertilizing capacity of semen specimens as characterized by ABO-secretor phenotype of the donor is affected by cooling and freezing. Thus, in summary, the effect of producing large numbers of offspring by means of frozen preserved semen would be to alter the "natural" balance pertaining to this genetic system.

In other words, donors who would be expected to father children in numbers which are directly proportionate to the number of times their semen specimens are used no longer father children in the expected proportions when their semen has been frozen. Those donors whose specimens display significantly greater freezability father a significantly greater proportion of offspring than would be expected on the basis of their normal fertility histories. Inevitably, this type of discrepancy in fertility as introduced by freeze preservation treatment will decidedly bias the number of offspring attributable to given donors.

It is clear from work already reported, on the other hand, that the fertilizing capacity of semen from normal donors is radically altered after having undergone freezing preservation treatment; e.g., spermatozoa aged during storage fertilize a significantly smaller proportion of eggs than fresh spermatozoa when they compete directly. Apparently this aberrance is not due to lack of fertilizing ability, embryonic losses, failure of capacitation, or lack of transport on the part of the aged spermatozoa.[12] Furthermore, it is interesting that in our experience[1] a conception has never been achieved when the frozen-preserved specimen had been administered more than 24 hours prior to ovulation as determined by basal body temperature charts.

Thus, there are two modes of inadvertent selection as described here: (1) selection operating upon the spermatozoa of a given ejaculate and (2) selection operating between donors. The effects of these two types of selection with respect to an entire population will be strictly dependent, in the final analysis, upon the proportion of all live births which is due to insemination with freeze-preserved semen. Although there is no means of keeping track of the multitudes of genes which would be affected in this way, we believe that it is incumbent upon us to demonstrate at least the order of magnitude within which we would expect significant changes to take place. It should be emphasized that the type of genetic effect we are discussing here is strictly statistical; there is absolutely no evidence indicating any increases in abortion, in congenital defects, or in any pathology as a result of using frozen semen. But our immediate success in individual cases with respect to the soundness of offspring should not prevent us from investigating the potential for population changes which are entailed by this clinical method.

The last area of discussion is intended to introduce ideas and problems which cannot be implemented or solved, in terms either of present-day scientific capabilities, or of contemporary medical ethics. In addition, it is not clear that the "enlightened" components of modern Western societies would find our subsequent proposals immediately acceptable or even deserving of serious consideration.

One topic is the role of cryogenics in systemic eugenics. The other is the role of frozen semen as an agent of selection. The first topic concerns an application of semen banks as a mechanism for positive eugenics; the second topic may be described by a possible "side effect" anticipated from the systematic and widespread use of frozen human semen.

In general, there are two methods whereby human spermatozoa can be readily manipulated for purposes of positive use. One method might be called simply, genetic engineering. In the most general terms, it implies the application of a biologic technique in such a way that the average genetically effective properties of sperm population are changed.

This type of genetic control applied to human gametes has never been attempted. Our interest in genetic engineering as applied to germ cells stems rather from the fact that while very large efforts are being expended to facilitate the genetic engineering which is required to correct somatic defects, it does not seem to have occurred to most workers that the ultimate success of genetic engineering techniques will depend upon the heritability of the genetic improvements which are made.

The second type of positive eugenics which can be applied to human germinal material, and which also depends upon the widespread availability of frozen preserved semen, has already been introduced and ably explained by H. J. Muller, the originator of the concept of parental choice. A quotation from a recent paper of Muller's follows:

"The main thesis I wish to uphold . . . is the following. For any group of people who have a rational attitude toward matters of reproduction, and who also have a genuine sense of their own responsibility, the means exist right now of achieving a much greater, speedier, and more significant genetic improvement of the population, by the use of selection, than could be effected by the most sophisticated methods of treatment of the genetic material that might be available in the Twenty-first Century. The obstacle to carrying out such an improvement by selection are psychological ones, based on antiquated traditions from which we can emancipate ourselves, but the obstacles to doing so by treatment of the genetic material are substantive ones, rooted in the inherent difficulties of the physical chemical situation.

"The proposed mode of procedure is to establish banks of stored spermatozoa, eventually ample banks, derived from persons of very diverse types, but including as far as possible those whose lives have given evidence of outstanding gifts of mind, merits of disposition and character, or physical fitness." We may note that there is some evidence that genes which determine such favorable characteristics have, on the whole, a significant amount of dominance. "From these germinal stores, couples would have the privilege of selecting such material, for the engendering of children of their own families, as appeared to them to afford the greatest promise of endowing their children with the kind of hereditary constitution that came nearest to their own ideals."[9]

While this concept is self-explanatory, we should note in addition, that this proposition is logically entailed by the current practice of heredity counseling, although the present status of such counseling constitutes a negative rather than positive eugenics.

The use of frozen preserved human spermatozoa for homologous or husband and heterologous or donor insemination has again been demonstrated with reasonable success. The major purpose of this paper, however, has been to indicate several points which require serious thought and serious research effort if the technique of semen freezing preservation is to be applied to the full extent of its potential. These are the following: first, that the cryobiology of human semen must be developed in its own right; human spermatozoa are biologically unique and their clinical application presents problems which are ethically and culturally unique. For this reason, the techniques which have been developed in the preservation of other mammalian gametes should not continue to determine the direction of our research in the problems of human biology.

Second, the systematic development of semen banks will facilitate an enormous amount of research which is immediately feasible and which is immediately necessary if the logic of this system is to be exploited fully. Third, we believe that the consequences of that widespread application of this technique will prove to be entirely determined within the basic tenet of biology, i.e., that of natural selection. This component of natural selection will, however, be entirely the responsibility of thoughtful scientists and clinicians who promote the use of this technique, whether for purposes of treating infertility or for purposes of genetic engineering or for purposes of positive eugenics.

References

1. Ackerman, D. R.: Fertil. & Steril. **19:** 123, 1968.
2. Ackerman, D. R.: Internat. J. Fertil. **13:** 220, 1968.
3. Behrman, S. J.: Techniques of Artificial Insemination, *in* Behrman, S. J., and Kistner, Robert W., editors: Progress in Infertility, Boston, 1968, Little, Brown & Company, pp. 717-730.
4. Behrman, S. J., and Sawada, Y.: Fertil. & Steril. **17:** 457, 1966.
5. Bunge, R. G., Keettel, W. C., and Sherman, J. K.: Fertil. & Steril. **5:** 520, 1954.

6. Fernandez-Cano, L., Menkin, M. F., Garcia, C., and Rock, J.: Fertil. & Steril. **15:** 390, 1964.
7. Keettel, W. C., Bunge, R. G., Bradbury, J. T., and Nelson, W. O.: J. A. M. A. **160:** 102, 1956.
8. Mantegazza, J.: R. C. I. Lombardo **3:** 183, 1866.
9. Muller, H. J.: Human Genetic Betterment, *in* Sonneborn, T. M., editor: The Control of Human Heredity and Evolution, New York, 1965, The Macmillan Company, pp. 100-122.
10. Osborn, Frederick: The Future of Human Heredity, New York, 1968, Weybright & Talley.
11. Perloff, W. H., Steinberger, E., and Sherman, J. K.: Fertil. & Steril. **15:** 501, 1964.
12. Roche, J. F., Dziuk, P. J., and Lodge, J. P.: J. Reprod. & Fertil. **16:** 155, 1968.
13. Sawada, Y., Ackerman, D., and Behrman, S. J.: Fertil. & Steril. **18:** 775, 1967.

21. Should Man Control His Genetic Future?

DONALD HUISINGH

In recent years, many scientists have begun to emerge from the ivory towers of pure scientific research and to become actively engaged in discussions with nonscientists. The scientists are beginning to realize their ethical responsibilities and obligations. Their concern has developed most rapidly since the discovery of the awesome power of atomic energy. This event has emphasized the need for scientists to inform their fellow citizens about the findings of science and the implications they may have for their lives and those of future generations.

Today scientists are speaking to representatives from all walks of life, including politicians, theologians, economists, and laymen in general. As concrete examples, three separate symposia dealing with "Man and His Future" have been held within the last four years.[1] The speaker lists were comprised primarily of physical and biological scientists.

None of the speakers claimed to have final answers about what direction man should take in the future. Most of them indicated various alternatives and discussed the probable results, but few grappled seriously with the quandaries that are likely to result in the pursuance of any particular course.

It is not surprising that most scientists are reticent to speak about the moral and ethical considerations of their work. They have tended to relegate religion to certain discreet times and places in their lives and to do the same with their science. Thus, few have had to grapple earnestly with the fundamental moral and ethical problems their work may raise. Furthermore, more scientists (I for one) came through undergraduate and graduate training in the physical and biological sciences with little formal experience in the social sciences. Scientists also appreciate the necessity to specialize and are aware of the pitfalls of speaking beyond their specialty, so they have tended to shy away from the territory of the moralist and the ethicist.

I feel uncomfortable in the role I try to fulfill in this paper: to write about the "Ethical Issues of Genetic Manipulation." I am not an ethicist, nor am I primarily a genetic specialist. In what follows, however, I will first attempt to sketch briefly some of the alternatives science has made

Reproduced by permission from *Zygon,* 1969, Volume 4, 188-199.

or is likely to make available to man to enable him to manipulate and direct the future of the human race. Second, I will discuss the problems involved in employing some of these alternative means and suggest tentative guidelines in their development and application.

POSSIBLE APPROACHES TO GENETIC MANIPULATION

There are three main categories of proposed approaches to genetic manipulation. They are: (1) euphenic engineering, (2) genetic engineering, and (3) eugenic engineering.

By *euphenic engineering,* Lederberg[2] refers to the modification or control of expression of the existing genetic information (genes) of an organism so as to lead to a desirable physical appearance (phenotype).

Genetic engineering is defined as the change of undesirable genes to more desirable forms by a process of directed mutation.

Eugenic engineering involves the selection and recombination of genes already existing in the "gene pool" of a population. The term eugenics was originally coined by Sir Francis Galton in 1883 to designate his aspiration to improve the human race by scientific breeding.[3] The word is derived from the Greek root, *eugenes,* which means "well born."

Euphenic engineering. Euphenic engineering in its simplest forms already is common practice. For example, lack of the capacity of an individual to produce insulin results in a disorder called diabetes. The expression of this genetic abnormality can be prevented by regular injections of insulin. Similarly, normal blood constituents such as gamma globulin now are supplied routinely to individuals who do not have the genetic information to synthesize these necessary blood components. Two other genetic defects lead to mental retardation because of the accumulation of harmful metabolic products. The diseases, phenylketonuria and galactosemia, result from an individual's inability to utilize the amino acid phenylalanine and sugar galactose, respectively. These diseases do not develop if the afflicted individual's intake of these molecules is restricted by careful control of his diet.

In these examples, expression of available genetic information was manipulated so as to minimize deleterious effects. As the factors which regulate and control gene action are more thoroughly understood, it is very likely that many other types of euphenic engineering will be possible. Suggestions of what the future may hold are evident by the following examples. It has been found that an injection of the anterior pituitary growth hormone into developing rats increased their brain size by 76 percent and increased their capacity to learn by an equivalent amount. There is one report of a similar response in a human child that received an injection of this hormone in its fourth month of fetal development. There is also a report of work being done in South Africa in which pregnant women are placed in decompression chambers for varying periods of time. The children which result are said to be superior to their siblings in intellectual capabilities.

It will be only a matter of time before many additional manipulations will be feasible, especially as we learn selectively to switch on or off at will the action of desirable or undesirable genes at specified periods in a person's life. The possibility of controlling the realization of the hereditary potential of the individual is impressive.

Genetic engineering. Genetic engineering is in its infancy in its applications to humans, but information already available suggests at least three possible approaches: (1) transduction, the virus-mediated transfer from one cell to another of genetic material; (2) transformation, the incorporation of a segment of DNA from one cell into the genetic material of another cell; (3) directed induction of mutations of specific places on the chromosomes (gene loci).

An example of transduction in humans was reported recently by Rogers of the Oak Ridge National Laboratory. He showed that the Shope virus, which causes tumors in rabbits, also induces the synthesis of a distinctive form of the enzyme arginase which lowers the concentration of the amino acid arginine in the rabbit's blood. Dr. Rogers wondered if this virus would also lower the arginine concentration in human blood. Because one may not infect human beings with animal viruses for experimental purposes, he had to get an answer to this question indirectly. He compared the blood of a number of people who had worked with, and therefore been exposed to, the Shope virus with the blood of randomly selected individuals as controls. He found that many of the researchers working with the virus were carrying "virus genetic information." They had lower arginine levels than controls and had specific antibodies against the distinctive form of arginase, indicating that the virus DNA had supplied the information for the synthesis.

The Shope virus, Rogers suggests, is a harmless "passenger" virus in these people. It is possible that there are other such viruses. Perhaps some of them carry genes that would be useful in the treatment of genetic diseases. It is conceivable that a harmless virus might even be utilized as a vector for specific information in the form of tailor-made DNA that could be attached to the virus and transferred by the process of transduction.

Szybalska of the McArdle Cancer Laboratory at the University of Wisconsin has reported that human cells in tissue culture can be transformed.[4] She found that the genetic ability to synthesize isosinic acid pyrophosphorylase could be transferred to cells that lacked this capacity by the application of DNA containing the appropriate genetic information.

Bentley Glass[5] in commenting on Szybalska's work has stated: "It may be feasible and possible in the near future to treat a germ cell defective in some gene with DNA from one known to be sound in that respect. By so doing it may be possible to improve the genetic content of the individual's reproductive cells and hopefully improve the performance of his progeny." This may be feasible, but not necessarily either advisable or wise. Within just a few years, however, we must decide whether to permit such engineering of human reproduction.

Tatum believes that "genetic engineering" by directed mutation can be seen as a possibility in humans. In microorganisms we already are learning techniques for producing mutations in a nonrandom fashion by the use of chemical mutagens such as nitrous acid and synthetic molecules related to nucleic-acid bases. These latter analogues are incorporated into DNA and upset the replicative process so as to cause the replacement of the original natural base by another one—thus producing a mutation.

Another potential approach to directed mutation is through the synthesis in the laboratory of a desired molecule of DNA. This tailored DNA molecule, if it can be isolated in pure form from an organism or cell, can probably be replicated by already known enzymatic processes to any needed quantity. This new or modified gene can then be introduced into the mammalian cell in culture as in bacterial transformation.

Eugenic engineering. Muller very pointedly says that while genetic engineering may have some applications in the future, it will be a long time before many of the technical difficulties are removed and the methods will be applicable to a sizable segment of the population.[6] Further, he indicates that euphenic engineering, while it may be extremely beneficial to the individual, does no good for the human race as a whole. On the contrary, by keeping a genetically defective person alive and allowing him to reproduce, we are increasing the frequency of deleterious genes in the population. Muller suggests, therefore, that we ought to employ a technique which is already possible: eugenic engineering through choice of desirable germ plasm.

Immediately, the term eugenics elicits a negative response on the part of many people, because to them eugenics and racism are synonymous. To others, eugenics means voluntary or mandatory sterilization of individuals who carry certain genetic defects. This latter approach to eugenics has been termed negative eugenics. Though such responses are understandable if the negative point of view of eugenics is maintained, Muller observes that they are readily modified by sincere thinking individuals if positive eugenics and the positive point of view are considered.

What is meant by positive eugenics? Muller develops his arguments for the application of artificial insemination with selected germinal material as a technique in positive eugenics. He says, "For any group of people who have a rational attitude towards matters of reproduction, and who also have a genuine sense of their own responsibility to the next and subsequent generations, the means exist right now of achieving a much greater, speedier, and more significant genetic improvement of the population, by the use of germinal selection, than could be effected by the most sophisticated methods of treatment of the genetic material that might be available in the twenty-first century."

The idea of artificial insemination per se is not new, nor is it completely objectionable. In the United States in 1962, more than 10,000 children were "fathered" by this method.[7] In most of those cases, the husband was either sterile or was carrying some genetic defect. The seminal donors were chosen by the doctor from men with body build and other morphological features similar to the husband, so that the resulting progeny would pass as the natural offspring of the legal family. The donor's anonymity was maintained to avoid paternity suits. According to Muller, we have among such couples many who would be happy to play a role in the decision of what germinal material is to be employed. "We are thus missing a golden opportunity to begin to consciously improve the genetic complement of the human race," Muller contends. According to him, intelligent germinal choice ought to be encouraged as the most effective way of rapidly achieving evolutionary improvement of the human race. Therefore, semen banks should be established, and the husband and wife ought to be permitted to select semen from donors of highest proven physical, mental, emotional, and moral traits. In order for a sound judgment to be made of the genetic potential of an individual donor, at least twenty years should be allowed to elapse after the donor's death before the deep-frozen semen is used. The men who earn enduring esteem can thus be called upon to reappear age after age through their preserved semen.

In addition to semen banks, it will soon be possible to store human ova as well. It is already possible to fertilize the human ovum in vitro and to implant the resulting embryo into the womb of a foster mother. It may also be possible in a few years to permit the embryo to develop to "normal" maturity in artificial glass wombs.

In fact, Dr. Daniele Petrucci of Bologna, Italy, has already done extensive experimental work with human embryos in vitro. Apparently, some technical difficulties still exist, because none of the embryos have lived beyond fifty-nine days.[8] There are also theological and legal difficulties; he was told by irate church officials and local legal authorities to discontinue this type of experimentation or be tried for murder. It takes little stretch of imagination to see that soon someone somewhere will make the necessary breakthrough and it will be technically possible to develop human beings in the laboratory from the sperm and eggs of any man or woman without restriction to time or place of the donor. In this way wide numbers of individuals could be produced by genetic selection from especially able parents. Further, as euphenic engineering progresses, it will be possible to nurture the developing embryos in different types of environments and thereby condition their mental constitutions. It will not be necessary for a woman to endure the discomfort and pain of carrying a child during the prenatal period. She would, of course, not get much psychological pleasure out of visiting the laboratory where her child was developing.

Does this sound too futuristic and too much like something taken from Huxley's *Brave New World?* Many people do not think so. In fact, in September 1965, the president of the American Chemical Society, Dr. Charles C. Price of the University of Pennsylvania, urged at the society's national meeting in Atlantic City that the United States make "creation of life" in the laboratory a national goal.[9] He was speaking of creation of life *de novo* from simple inorganic and organic molecules, a feat far more complicated and difficult than merely growing an embryo to maturity in vitro.

I have tried to indicate some of the types of possibilities that are or may be feasibly applied in directing the future of man. Many questions and problems present themselves. Guidelines are needed to help the scientist choose what types of research he ought to be engaged in, to aid the technologist to select the techniques he should make available to the population at large, and to enable the politician to cope with the social, legal, and political problems which will arise as these techniques are used. With this "Biological Bomb" already about to explode, the need to face the complexity of the problems involved takes on acute, do-it-now urgency.

PROBLEMS AND GUIDELINES

The possible approaches to genetic manipulation we have discussed raise many ethical questions, including: (1) What is the essence of human life? (2) Is the human body a sacred vessel of man's soul and spirit, or is he merely at that position in biological evolution to know that he is a part of evolution and can do something about his own future evolution? (3) What absolute human values are we eager to retain? (4) What values are only relative in a particular sociological, theological, and political framework and as such should change with future evolutionary changes? (5) What are man's biological rights and responsibilities as individuals and as members of the species *Homo sapiens?*

If we accept the possibility of improving man by germinal and oval selection, the following questions arise: (6) Are there any truly ideal genotypes? (7) What are objective criteria by which they can be selected? (8) Who singly or collectively could objectively select individuals who fulfill these criteria once they are agreed upon? Since the eugenic approach would necessarily be a long-term project, it would have to have built-in mechanisms to ensure that the goals and objectives did not change with every new generation. This would necessitate the development of breeding plans. (9) Could such plans for the population be pursued without at the same time taking away much of the freedom of the individual? (10) Should the individual's freedoms and rights be secondary to the supposed good of the human race as a whole?

Let's take a look at what might happen if we subject man to a program of planned eugenics. William Shockley in a discussion of this subject stated, "I believe the difficulty with planned eugenics is that we are forced to think of ourselves and other people as being not solely warm, living human beings with whom we can establish personal relationships, but as *objects* which can be thought of and dealt with statistically and analytically."[10] He goes on to say, "My own reaction reminded me of a quotation expressing the same feelings in T. S. Eliot's *The Cocktail Party*":

> Nobody likes to be left with a mystery.
> But there's more to it than that. There's a loss of
> personality;
> Or rather, you've lost touch with the person
> You thought you were. You no longer feel quite human.
> You're suddenly reduced to the status of an object—
> A living object, but no longer a person.
> It's always happening, because one is an object
> As well as a person. But we forget about it
> As quickly as we can. When you've dressed for a party
> And are going downstairs, with everything about you
> Arranged to support you in the role you have chosen,
> Then sometimes, when you come to the bottom step
> There is one step more than your feet expected
> And you come down with a jolt. Just for a moment
> You have the experience of being an object
> At the mercy of a malevolent staircase.
> Or, take a surgical operation.
> In consultation with the doctor and the surgeon,
> In going to bed in the nursing home,
> In talking to the matron, you are still the subject,
> The centre of reality. But, stretched on the table,
> You are a piece of furniture in a repair shop
> For those who surround you, the masked actors;
> All there is of you is your body
> And the 'you' is withdrawn.[11]

Do we dare withdraw the "you-ness" from human beings? Do we have the option to treat man as a manipulable object or is he to be treated as an inviolable individual at the center of reality? Is there something here that we must strive to retain? Is it possible that in the process of attempting to call the shots for human evolution, one will destroy those attributes that make him human? Muller says, "No, for by selection of individual sires who have demonstrated a genuine warmth of fellow-feeling, a cooperative disposition, a depth and breadth of intellectual capacity, moral courage, an appreciation of nature and of art, a healthful, vigorous constitution, and highly developed physical tolerances and aptitudes; the progeny of such progenitors are bound to be more human, not less."[12]

The question of our ability to select objectively such qualities looms as a great obstacle; but the problem of finding anyone with all of these attributes in desirable proportions is even more formidable. Besides that, it is believed that everyone carries an average of four to ten recessive lethal genes which express themselves only in the homozygous condition or only in the presence of certain modifier genes or under certain environmental conditions. Even the best phenotype may have these genes lying hidden.

Let's assume, however, that we found outstanding individuals of the types desired. What is the probability of improving the heritage of the human race appreciably? To answer this, I quote from Bentley Glass's article, "Human Heredity and the Ethics of Tomorrow":

> The fertilized egg contains 46 chromosomes, 23 of them inherited from the egg and 23 from the sperm. The number of different genotypes that might be present in a single fertilized egg, if there were only 23 differences between the genes in the two sets of chromosomes in the father, and 23 other differences between the genes in the two sets of chromosomes in the mother, i.e., one difference per pair of chromosomes, would be $(2^{23})^2$. That is to say, the mother could potentially produce 2^{23} or 8, 388, 608 genetically different sorts of eggs, and the father an equal number of sperms with different genotypes. Hence there is the possibility through random fertilization of nearly 70 trillion genotypes of offspring. That would amount to about 2,300 times the present population of the entire world. This means that the variety of human genotypes is essentially inexhaustible and that there is only an infinitesimal chance that any two persons will be identical in all genetic respects with the exception of identical twins, triplets, etc.[13]

Besides this, many of the desirable attributes of man are inherited not as single genes but as multiple genes.

Thus, even if we were capable of selecting outstanding semen and ova donors, a large degree of variability would be expected in the progeny. There would, however, be a large number of genes held in common by the offspring. Not only would it be undesirable sociologically to have large numbers of humans of very similar genotypes, but biologically it may even prove to be disastrous, because nature puts a premium on variability. Let me cite two examples of what could happen if homogeneity were achieved in a large segment of the human population.

Wheat breeders have selected, crossed, backcrossed, and selected again for desirable agronomic qualities in wheat. A few years ago, they came out with a variety which they said had tremendous genetic potential for productivity and also carried a high degree of disease resistance. In a few years, thousands of acres of America's rich wheat lands were planted with this variety of wheat. But, in 1952, a new race of wheat stem rust appeared which completely overcame the disease resistance of this variety. Two years later only a few acres of this wheat were to be found anywhere in the country. That variety of wheat was fine in the old environment but not in the new one; so too is the possibility with man. New strains of bacteria, fungi, and viruses are arising all the time. If there were a large number of people who held many genes in common, they could all rapidly succumb to a new strain of microorganism that was pathogenic against those genes.

An example from the genetics of fruit flies also is relevant here. In a certain strain the female flies have been shown to pass on through their eggs a virus which infects the developing young. This virus makes the individuals susceptible to CO_2 and has been called the CO_2 lethality virus. Similar transovarian transmission of viruses in humans is possible, and if the ova of several individuals were widely used, the likely results are obvious.

This discussion of the application of eugenics for human betterment has been based upon the following assumptions: (1) we could objectively agree what qualities to select for, (2) we could quantify and select these qualities, and (3) genetics is the most important factor in determining that an individual have or develop the desirable traits listed earlier.

Without going too deeply into the nature versus nurture argument, I would like to cite the statements of two noted authorities. Nobel laureate Dr. Francis H. C. Crick, physical biologist at Cambridge University and winner of the Nobel Prize in 1962 for his contribution to our understanding of the physical arrangement of DNA, is reported to have said, "Humans probably will not be improved or altered by genetic manipulation in the future; education and environment are more important than genetics."[14]

Dr. Bentley Glass, in a similar tone, says:

> Modern man has been on the earth for an immense stretch of time, at least 40,000 years, and maybe several hundred thousand, without much change in his skeletal

anatomy. We are therefore justified, I think, in regarding all his tremendous human advance in culture and civilization, in material power and relative understanding of nature, as having occurred with little if any, genetic change. The great advances made by modern man, therefore, reflect no change in his biological heritage but represent a new phenomenon, the advent of cultural transmission, the accumulation of knowledge and its transfer from one generation to the next. Equal opportunity must be coupled with freedom of the individual if it is to lead to fullest development of the potential of the genotype.[15]

Thus two outstanding experts in the area of genetics seem to say that it is not desirable, nor is it likely to be fruitful, to involve human beings in Muller's grand genetic experiment. I concur with that conclusion. Most people never approach a realization of their genetic potential even for short periods of time because their society has not provided the educational opportunities for them to develop fully their intellectual capabilities. In additional cases the sociological mores, the political machines, and the theological institutions have erected further barriers to the individual's progress. In short, their individuality, personality, and humanity have never been developed. We ought to concentrate on maximizing the nurture of every individual so that he more nearly realizes his existing genetic potential, and move very cautiously into the area of germinal selection experiments.

Much more could be stated in summary about other possible approaches to the future of man. For example, it seems likely that negative eugenics should be continued, but it poses the ethical problem of selection of some individuals as not being fit to reproduce. In some extreme cases few people would disagree that some individuals do not have the right to reproduce because of their load of genetic defects. However, very few people would agree where the cutoff point should be.

I anticipate that genetic engineering will be found to be helpful in modifying the genetic information of individuals, but it is not likely to be a significant factor in the human population as a whole. On the other hand, progress in euphenic engineering and the provision of favorable conditions for human development and self-actualization are likely to be the most significant ways by which man will direct his own evolution. There will be many facets of the "scientific engineering" of man which will pose serious ethical problems. The politicians, lawyers, theologians, social scientists, and the general public must all be informed of the alternatives which will be available to them. They must be informed of the possible benefits and problems in order to be in a position to make intelligent use of the new tools science will provide.

What general guide will all these people use in deciding whether a particular type of research should be engaged in or whether a particular practice should be condoned and used? I would like to paraphrase what John Baillie has to say in his little book, *Natural Science and the Spiritual Life.* The future of man is secure only as long as the virtues of humility, tolerance, and impartiality are retained as absolute standards. Within this framework we should use scientific advances as tools to serve society.[16]

The time ahead is uncharted. No one has been there, so there are no experts. Each of us whose body and brain may be modified or whose descendant's characteristics may be predetermined has a vast personal stake in the outcome. We can help to insure that good will be done only by looking to it ourselves. We must be careful to retain the individuality of the individual and the personality of the person, or else the humanity of the human may be lost.

Notes

1. See J. D. Roslansky, ed., *Genetics and the Future of Man* (New York: Appleton-Century-Crofts, 1966); T. M. Sonneborn, ed., *The Control of Human Heredity and Evolution* (New York: Macmillan Co., 1965); and G. Wolstenholme, ed., *Man and His Future* (Boston: Little, Brown & Co., 1963).
2. See Wolstenholme, pp. 263-73.
3. Anram Scheinfeld, *The New You and Heredity* (Philadelphia: J. B. Lippincott Co., 1950), p. 679.
4. Elizabeth H. Szybalska and W. Szybalska, "Genetics of Human Cell Lines. IV. DNA—Medicated Heritable Transformation of a Biochemical Trait," *Proceedings of the National Academy of Sciences* 48 (1962): 2026-34.
5. C. Stern, *Principles of Human Genetics* (San Francisco: W. H. Freeman & Co., 1960), p. 753.
6. See Sonneborn, pp. 100-122.
7. See Wolstenholme, pp. 258-62.
8. "Control of Life. I. Exploration of Prenativity," *Life,* September

10, 1965, pp. 60-79; "Control of Life. II. Gift of Life from the Dead," *Life,* September 17, 1965, pp. 78-88; "Control of Life. III. Spare Parts for People," *Life,* September 24, 1965, pp. 66-84; and "Control of Life. IV. The New Man—What Will He Be Like?" *Life,* October 1, 1965, pp. 94-111.

9. "Control of Life. IV. The New Man—What Will He Be Like?" *Life,* October 1, 1965, p. 94.
10. See Roslansky (n. 1, above), pp. 96-97.
11. T. S. Eliot, *The Cocktail Party* (New York: Harcourt, Brace & World, 1950), pp. 29-30. Reprinted by permission of Harcourt, Brace & World, Inc.
12. See Wolstenholme (n. 1, above), pp. 247-62.
13. Bentley Glass, *Science and Ethical Values* (Chapel Hill: University of North Carolina Press, 1965), p. 101.
14. "Little Change Likely," *Industrial Research* (March 1967), p. 36.
15. Bentley Glass, "The Ethical Basis of Science," *Science* 150 (1965): 1254-61.
16. John Baillie, *Natural Science and the Spiritual Life* (New York: Charles Scribner's Sons, 1952), pp. 34-43.

The Final Question

Do humans have the genetic constitution to control the future evolution of humans?